THE
ECO HERO
HANDBOOK

THE
ECO HERO
HANDBOOK

SIMPLE **SOLUTIONS** TO
TACKLE ECO-ANXIETY

TESSA WARDLEY

First published in the UK and North America in 2021 by

Ivy Press

An imprint of The Quarto Group
The Old Brewery, 6 Blundell Street
London N7 9BH, United Kingdom
T (0)20 7700 6700
www.QuartoKnows.com

British Library Cataloguing-in-Publication Data
A catalogue record for this book is available from the British Library.

ISBN: 978-0-7112-5463-3

This book was conceived, designed and produced by

Ivy Press

58 West Street, Brighton BN1 2RA, UK

Publisher **David Breuer**
Editorial Director **Tom Kitch**
Art Director **James Lawrence**
Commissioning Editor **Niamh Jones**
Project Editor **Stephanie Evans**
Design Manager **Anna Stevens**
Designer **JC Lanaway**
Illustrator **Melvyn Evans**

Printed in China

10 9 8 7 6 5 4 3 2 1

Contents

Introduction

Every day in the media we see and hear that our world is in peril: the sixth extinction, melting ice caps, sea level and temperature rises, increased desertification and ever-more frequent and devastating extreme weather events.

Fears of planetary fragility were ignited in 1958 when scientists reported rising levels of carbon dioxide globally, and in 1962 American biologist Rachel Carson highlighted the impacts of using deadly pesticides in her book *Silent Spring*. Evidence grew of the long-term effects of chemical pollution, burning fossil fuels, depleting natural habitats and using non-renewable resources. The hum of concern persisted in the 1970s and 80s; Greenpeace was founded in 1971 but political will stumbled far behind in the intervening years. Al Gore and David Attenborough voiced the risks our current activities and economic focus pose for our planet and in 2015 it seemed consensus on the challenges and solutions to the climate crisis had been reached when United Nations countries signed the Paris Agreement at the UN Framework Convention on Climate Change, but the general feeling of unease spiked with the publication of two devastating reports; one from the Intergovernmental Panel on Climate Change (IPCC) in 2018 and another from the UN's Intergovernmental Science-Policy Platform on Biodiversity and Ecosystem Services (IPBES) in 2019.

The reports showed that if we continue on our current trajectory, uncontrollable climate change is just a few years away and biodiversity on Earth will decline at a dizzying rate, eroding, to quote the IPBES Chair Robert Watson 'the very foundations of our economies, livelihoods, food security, health and quality of life worldwide'. That message hit many of us hard. Collective anxiety exploded, spurred on by the catastrophic news contained in the reports and the perceived lack of political action. Feelings of grief, rage, despair, guilt and shame led to mass protests and demands. Youth and social action such as school climate strikes inspired by Greta Thunberg, Extinction Rebellion and other public protests around the world are demanding action from our leaders and policy makers. Governments respond and, sometimes, appropriate announcements are made but implementation lags dangerously behind. Signatories to the Paris Agreement are projected to miss targets in 2025 and 2030 while President Trump subsequently withdrew the USA from the agreement entirely.

At a personal level, 'eco-anxiety' is actually a normal and healthy human response to the facts. We must process our emotions; we cannot live in denial but we must live with ourselves. Conscious actions to take control of CO_2 emissions as individuals on a daily basis can help our mental state. Being informed and getting involved can help us to eat, travel and live in less consumptive ways.

It is tempting to feel insignificant in our actions, that we cannot influence the direction in which the world is heading, but our collective power is great. The direct effect of personal actions may not be immediately obvious but the ripples they send out and how they multiply may well be. This is what is known as the butterfly effect.

When Greta Thunberg decided not to go to school in August 2018, she never imagined that within six months she would be talking to global leaders, alongside David Attenborough, at the World Economic Forum in Davos, Switzerland. We too can make an impact; as Greta put it, no one is too small to make a difference.

This book is a practical guide to help us all make better choices, to maximize our influence despite the global nature of the challenge, and get us all on the right track. The actions that are within our control, to change the world for the better, are organized here into sections – in the home, out and about, travelling, at work or school, on holiday, and food and shopping – and each issue comes with a solution that really does make a difference.

1.

INDOORS

Our homes are our havens. When we create a home, we have freedom to express ourselves as we develop a comfortable space in which to relax, at one with ourselves and our surroundings.

In our homes we have the opportunity to experiment with green living and find the best solutions for us and for the planet. What better way to demonstrate our green credentials, to illustrate our values than through the way we decorate, heat and source the materials for where we live; our shelter and our sanctuary?

The changes we make at home support our physical and our mental wellbeing. They reassure and comfort us as we are able to be the change we want to see. With careful consideration of the available options we can make choices that influence our whole approach to life; we can assert our belief systems to demonstrate to ourselves and the world that we are ecologically aware and socially responsible citizens. By making informed decisions it is possible to minimize our emissions and drive a positive effect on the natural world.

How can I keep my house clean but be low impact?

ISSUE

Why do we use hazardous chemicals packed in single-use plastic bottles to clean our homes? They're bad for our health and disastrous for the environment because they end up in rivers and oceans, poisoning plants and animals and disrupting the balance of nature.

SOLUTION All evidence suggests that for everyday cleaning ordinary soap, warm water and plain detergent are as good as antibacterial or antimicrobial products (unless you have a specific health problem). These products can kill off the good bacteria on our skin and contribute to the problem of antibiotic-resistant bacteria.[1]

Microfibre cloths can be used with just water. Water is a universal solvent and the microfibre design means that water fills the spaces in the fibres, attracting and holding dirt, grime, mould – even bacteria. Simply wash the cloth to release the particles, and reuse. Microfibre cloths remove more than 99 per cent of bacteria using just water, making them the perfect option for a less toxic cleaning routine. Look out for natural microfibres; most on the market are petroleum-based.

For a deeper clean, vinegar is an effective multipurpose disinfectant for most surfaces. Mix white vinegar and water 50:50 in a spray bottle for use on most surfaces. For a fresh smell add a few drops of citrus essential oil or a couple of teaspoons of orange or lemon juice. Undiluted white vinegar mixed with bicarbonate of soda will descale kettles and toilets. Leave overnight then rinse or flush in the morning.

If you need to deep clean, buy products that are biodegradable, non-toxic and free from parabens, phosphate, sulphate, synthetic perfume and bleach. Choose a multipurpose cleaner; don't fall for the marketing ploy encouraging you to spend more money and fill your cupboard with plastic. Many natural products offer a refilling service or bulk-buy containers that you decant into a reusable spray bottle.

. . . .

Good to know . . .

• Disinfectants can cause resistant bacteria, mutagenic bi-products, kill helpful bacteria in treatments plants, affect sewage treatment performance and contaminate surface water

• Other effects include contributing to poor indoor air quality, cancers, reproductive and respiratory disorders, eye and skin irritation, and central nervous system impairment.[2]

* Here disinfectants can mean formaldehyde, glutaraldehyde, bleach, hydrogen peroxide and enzymatic cleaners.

Is my water usage affecting the planet and nature?

ISSUE

Globally, our use of water is unsustainable and the consequences are far-reaching. We see ecological function break down in the loss of riverside trees and migratory aquatic species, increases in algal blooms and invasive species, and decreases in bird populations.

SOLUTION The hydrological cycle – the continuous flow of water between the atmosphere, land and sea via precipitation and evaporation – is key to life, not simply for humans but the plants and animals that have adapted over millennia to the natural conditions. Being less wasteful with water benefits river systems and aquatic species, reduces the stress on plant, animal and bird populations, and makes us better able to respond to the consequences of climate change, such as droughts or forest fires.

Simple daily actions to reduce your consumption include keeping a jug of water in the fridge rather than running the tap for a cold drink, and turning off the tap while brushing your teeth – leave it flowing for 2 minutes and you waste up to 12 litres (more than 2½ gallons) of water. Wash fruit and veg in a bowl of water and use it for watering house plants. Use a bucket rather than a hosepipe for cleaning cars, dogs and watering your garden. Or collect rainwater and use that for free. Only run your washing machine and dishwasher as full loads and fix any leaks in your home. Take a shower (10 minutes max!) rather than a bath, which can use up to 80 litres (17½ gallons). If you do want a soak, run it less deep.

Globally daily average water consumption is **160 l. (35 gal)** per person but it varies widely by country:
- **900 l. (198 gal)** in Australia
- **750 l. (165 gal)** in Canada
- **600 l. (132 gal)** in the USA
- **150 l. (33 gal)** in the UK
- **70 l. (15 gal)** in China
- **40 l. (9 gal)** in Bangladesh.

. . . .

Good to know . . .
- Water covers 70% of the earth's surface but most is saline sea water or locked in ice caps; less than 1% is fresh water.
- Around 20% of the world's population is unable to access safe drinking water, whereas in developed cities water of drinking quality is used to flush our systems and water our gardens – only 1% of our drinking standard water is actually used for drinking.[3]

How do I use drains without harming the environment?

ISSUE

Wasting water is only part of the concern. Sewers can become blocked by huge 'fatbergs' (non-biodegradable stuff mixed up with congealed grease or cooking fat) and chemicals and other waste enter the water system. What can I safely put down the drain or toilet?

SOLUTION Any waste you put down the kitchen sink or house drain enters the sewerage system and undergoes sewage treatment. Once solids are removed and the sewage has been treated the effluent will ultimately be released onto land, into waterways or the sea.

To protect your own plumbing and the sewerage system there are things you should never put down the sink, toilet or your house drains. The biggest culprits are oils and fats which harden as they cool. Instead of pouring them down the sink let them cool before binning them, or mix with seeds or uncooked rice to make your own fatballs to attract birds to your garden (see page 47). Coffee grounds and eggshells are another no-no, so either compost or bin them too. Non-biodegradable items such as feminine hygiene products, cotton balls and pads, wipes and condoms can cause blockages and the only place for them is the bin.

To protect the environment, don't flush unused medication down the sink or toilet – once dissolved in water it cannot be removed by treatment works. Pharmaceutical drugs are damaging aquatic wildlife[4] and can end up back in our drinking water supplies. Don't use beauty products containing plastic microbeads – they are not removed by treatment and don't biodegrade; instead, they build up in the bodies of sea animals. Avoid buying cleaning products if they are not biodegradable or contain toxic chemicals or phosphates which can kill off on-site treatment systems and are hard to remove at treatment plants, meaning they will end up damaging the environment.

Outside of your home, on the street and kerbside, ditches and drains discharge directly onto land or into rivers or the sea with minimal treatment. They are designed to take away rainwater so make sure you never put any oil (whether from your kitchen, workshop or car), leftover paint, cigarette butts, dog waste or anything else down them – they collect enough nasties from road run-off; you don't want to be adding other polluting waste.

Are all plastics bad and can we ever do without them?

Plastic is everywhere. It litters our streets and countryside, and media images show it ends up in the bodies of marine life or forms islands in the oceans. Now I've read it's reached the deepest sea trenches and the most remote parts of Antarctica. Is any plastic use OK?

SOLUTION Plastics in general are bad for us and for the planet. Their carbon footprint is massive because most are derived from coal, natural gas and crude oil and use energy, often from fossil fuels, in their manufacture. As products they are not great either. Chemicals can leach out and are believed to account for declining sperm counts, higher incidence of some cancers and other negative impacts.[5] Recycling plastic is energy intensive too. Technically all plastic can be recycled but there are logistical challenges (see page 26). The resulting recycled product is low-grade, far more costly than landfill and not economically viable. Microplastics (see page 38) and biodegradable plastics (that break down rapidly into microplastics) are a particular problem – impossible to treat and easily ingested by wildlife. We can all avoid single-use plastics by reusing bags, taking our own cups when we buy coffee on the go, refilling the plastic we already have, and buying glass not plastic bottles.

There are arguments, though, for retaining plastics in some industries: being light, flexible, durable, transparent, inert and cheap, plastics reduce transportation costs and waste in our food supply chains. For example, food waste (see page 112) has ten times the environmental impact in terms of its carbon footprint than the use of plastic to prevent waste.[6] And it would be hard in the medical world to do without plastics – they are used in fine tubing, syringes, blood and saline bags, replacement joints, prosthetic limbs, optical lenses, pill casings and MRI scanners.

. . . .

Good to know . . .
- Globally we recycle under 20% of our plastic.
- Around 55% of plastics are used once and discarded.
- Of the top ten items that end up polluting the oceans, nine are disposable single-use plastics: cigarette butts, wrappers, bottles, bottle caps, bags, straws or stirrers, plastic or foam containers and lids.
- An estimated 8 million tonnes of plastic ends up in our oceans and seas annually, causing the deaths of thousands of marine mammals and birds each year, according to the UN Environment Programme.[7]

How can I reduce my overall energy use in the home?

ISSUE

Our energy use keeps going up. Today we use about three times as much as we did 50 years ago, and around a third of that energy is used directly in the home. We all need to look at ways to reduce our energy consumption at home.

SOLUTION Fifteen per cent of the planet's greenhouse gases come from heating our homes and there are many ways to reduce your energy consumption and save money as well. First, insulate your house – loft insulation pays for itself within a year (2–3 years if you're paying someone to install it) while wall insulation will take 5 years to pay back. Double or triple glazing doors and windows and excluding draughts can also mean lower energy bills and make your home more comfortable. And for every degree you turn down your thermostat, you save around 10 per cent of energy. But don't go too low; medical professionals recommend 18°C (64°F) as the minimum for people over 65 and those with medical conditions.

Do ensure all your lighting uses LED bulbs – they are a bit more expensive to buy but they last much longer (typically claiming 25–30 years) and use 90 per cent less energy than incandescent bulbs, so you will see savings within a couple of months.

Turn off everything when not in use or when you go out. Heating an empty house is obviously a waste but electronic appliances on standby are equally wasteful; your digital TV box and TV, multiroom speakers, routers, microwave, phone charger when not charging and especially personal computers and games consoles all use a lot of energy and that costs you money. So try a bit of *hygge*, snuggle up with jumpers and blankets and remember when you are going out, and particularly when you are going away, to turn all your appliances down or off.

. . . .

Good to know . . .

• Taking the global average, each person uses 59 kWh per day[8] – the same as a kettle and toaster heating non-stop day and night – or enough calories to feed 22 people for a day.[9]

• A European uses twice the global average; an American four times as much; an African only a fifth.[9] Within the day usage varies: asleep in an unheated house you are only using about 3% of the global average[10] but fly alone on a private jet and you are using in the order of 1,000 times the average.[11]

How can I cut the energy and water used by white goods?

ISSUE

Every large appliance requires energy and resources in production, in use and in disposal. We don't expect to replace our white goods often, so mainly our concern is to ensure that we are using these machines most efficiently to keep resources and energy-use to a minimum.

SOLUTION It's a good idea to read the instructions for your white goods – there may be energy- and water-saving settings you unaware of and tips on how and where to install them for best efficiency. For example, the space around free-standing rather than built-in fridge/freezers can allow waste heat to escape, cooling the appliance and saving up to 150 kg (330 lb) of CO_2 emissions a year.[12] White goods operate most efficiently on full loads so delay washing your dishes or clothes until you can fill the machine. And check out special features – some models have half-load and 'eco' settings that save water and electricity. If you can't fill your dishwater, then wash by hand. Wash your clothes at the lowest effective temperature because around 90 per cent of the energy used by washing machines is to heat the water. Modern detergents work at 30°C (86°F) so there's no need to wash at higher temperatures on a regular basis – the occasional hot wash of bedding (over 50°C/122°F) will kill off bacteria and bugs. A tumble dryer uses a lot of energy so only use if essential – hang your washing outdoors in the fresh air to dry whenever possible.

Turn off your white goods when they are not in use, or ask if you really need them all – many households keep an old fridge in the garage, humming away with a couple of beers and some ancient peas in it. Keeping your fridge door shut and running it at 4–5°C (about 40°F) is the most efficient for reducing food waste.

If a machine goes wrong, try to get it fixed before disposing of it. If you do need to replace it, look into recycling schemes through your local council, charities and other organizations. And when buying new white goods, in most countries they require an energy and water rating so you can choose one that minimizes consumption. This makes good environmental sense but also good financial sense. In the long run, paying a bit more for a better-quality white good – one that won't need replacing for many years, and that is designed to use less electricity and water – will invariably save you and the planet.

Is getting a smart meter a good idea?

I know that smart meters are said to be the next generation of electricity and gas meters. But I don't really understand what a smart meter is and why it should have a positive impact on the environment. Should I get one?

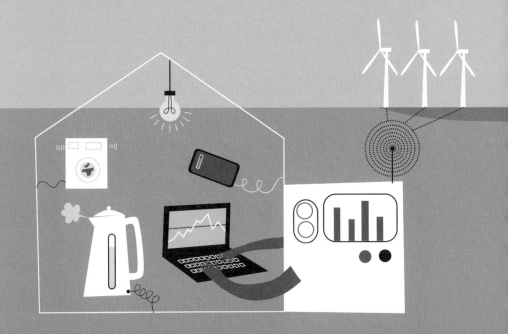

SOLUTION Like your traditional meter, a 'smart' meter measures your total energy use while also showing you how much you are using as you use it and how much it costs. It can also provide your historical usage, so you can compare. All this detail is displayed on a small device in your home, and it is relayed to your energy supplier, allowing them to check usage remotely and eliminate the need for meter readings. Tariff changes and bill adjustments can also be made immediately and remotely so that you won't be paying for energy you are not using.

With a smart meter you can track your usage and get immediate feedback on how much energy you are using with each appliance or activity. This allows you to change your behaviour, helping you to save carbon and money. In the UK 85 per cent of new smart-metered households altered their behaviour, which resulted in energy savings. More people are having smart meters installed and an estimated 24 per cent decrease in emissions by homes and businesses in the UK is expected by 2030.[13]

However, not all suppliers use smart meters, so you may not have the option to obtain one unless you switch. And because smart meters rely on mobile networks you need to be in an area with good reception. Finally, simply installing a smart meter isn't enough: it works by encouraging us to change our behaviour, reducing our energy consumption and emissions and saving money. If you can, get a smart meter and become part of the change we need to achieve net zero carbon.

Without smart meters, meeting climate change targets will be slower, harder and more expensive.[14]

· ·

Good to know . . .
• Globally, the number of smart meters is set to almost double from 665.1 million in 2017 to more than 1.2 billion by 2024. This surge in take-up is being driven primarily by the Asian market but also Europe, North America, Australia, New Zealand, Latin America and to a lesser extent Africa.[15]

What can I do to reduce my online impact?

ISSUE

Every second of every day people are browsing the web, checking and posting to social media, sending emails, watching videos and listening to music and podcasts – and it all uses electricity. What can I do to minimize my impact without switching off completely?

SOLUTION We think of the internet and the 'cloud' as a magical place where 'stuff happens'. The reality is warehouses bigger than aircraft hangars, filled with servers and circuit boards, burning electricity and pumping out heat. They connect via tens of thousands of miles of cables over the ocean floors. Our interactions with the cloud, web searches, emails and use of social media all add to greenhouse gases.

We can reduce our carbon footprints by increasing our energy efficiency. Start by using the right kit and keeping it for as long as possible: don't duplicate devices but choose the right one for the task. For example, it takes 75 per cent less material to manufacture a laptop than a desktop computer and a laptop consumes 70 per cent less energy when in use.[16] Similarly, for web browsing or processing emails, a tablet is more efficient than a laptop. Attaching by cable uses less energy than WI-FI and 4G or 5G, so if possible connect to screens, printers, networks and cloud storage via ethernet and USB cables. Use features that automatically put devices on standby mode and don't let batteries discharge completely. If a device has an optimization feature, use it to stop apps running in the background and to empty unnecessarily occupied memory space. Add your regular websites to 'favourites' and, if you know a website address, enter it rather than searching; it takes less energy than bringing up options. Similarly, more accurate searches help reduce the number of sites you view.[16] When emailing send hypertext links rather than attachments or compress the documents. Once dealt with, delete emails and empty your trash – it all helps.

. . . .

Good to know . . .

• The emissions generated by the global IT industry are equivalent to those of the global aviation industry, primarily because of the energy required to power data centres.

• Despite efficiency measures being developed all the time, emissions are expected to grow again by 2030.

• Internet traffic is growing exponentially. Every minute 570 websites are created, 340,000 tweets sent and a quarter of a million searches made on Google alone.[17]

How can I recycle packaging effectively?

There are so many rules and things to know about recycling; do I need to wash things out, should I leave lids on or off and should I crush or not? How can I make sure I don't get it wrong and cause an entire collection of recycling to be rerouted to landfill?

SOLUTION We need to reduce the amount of packaging used, but we also want to ensure that what we send for recycling is *actually* recycled. Here are some pointers.

Bottles, jars and cartons (plastic, glass or foil-lined cardboard) are all recyclable, but they must be completely empty and rinsed. This is important: residual food can attract small mammals and birds or contaminate otherwise recyclable materials. An overcontaminated load of recycling will be rerouted to landfill or incineration. Flatten plastic bottles and cartons and screw on any lids. Aluminium drinks and food cans, emptied and rinsed, can be recycled, and likewise aluminium tubes, for example from tomato purées – just squeeze them as empty as possible.

Pizza boxes (clean) go in with your cardboard and envelopes with windows can be recycled with your paper. Wrapping paper can be recycled unless it has glitter or plastic additives – check by scrunching the sheet: if it unfolds it contains plastic.

Fruit and veg punnets can be recycled in your kerbside collection but first remove any plastic film or cling film. These films – along with bubble wrap, magazines and carrier bags, shrink wrap on multipacks, and bread, cereal and frozen food bags – may be returned to collection points for specialist recycling, away from standard recycling system. Exclude compostable or biodegradable bags as they mess up the system. Black plastic used in some meat and fish trays and plant pots can confuse recycling systems so are generally not recyclable.

Remember to keep a recycling bin in the bathroom – only 50 per cent of bathroom plastics are recycled compared with 90 per cent in the kitchen. For products that can't go into kerbside recycling, such as dog and baby food pouches, contact lenses, crisp packets, and rubber gloves, new recycling options are opening up all the time, so keep an eye out in your local area for specialist recycling schemes.

As ever, try to throw out as little as possible and reuse wherever you can before things go in the bin. This is particularly important for items that cannot be recycled.

What can I do to limit the toxins I breathe in at home?

I'm concerned to read that levels of air pollutants indoors can be two to five times higher than outdoors. With volatile organic compounds, pesticides, chloroform, esters, ethers, dioxins, formaldehyde, PVC and radon in everyday products, how can I improve the air in my home?

ISSUE

SOLUTION It is a concern that toxins in the air we breathe are damaging our health, but there are some simple proportionate ways to prevent the build-up of these volatile toxins in our homes that will also help the wider environment.

A smoke-free home is an obvious start and, while you are at it, make it shoe-free – many pollutants come in on dirt particles, so use door mats to remove 60 per cent of dirt on entry, vacuum rugs and carpets regularly and mop hard floors with a microfibre mop (water alone is enough to pick up dust particles).

Avoid chemicals in insect-pest treatments and stick to natural products for cleaning, bath and beauty products. Synthetic fragrances in laundry detergents, room sprays, dryer sheets, fabric softeners or plug-in air fresheners are not really making your air fresher, they are putting toxic volatile organic compounds (VOCs) into your home – so cut them out. Simply cleaning with white vinegar and bicarbonate of soda with added lemon (see page 11) gives a fresh scent in the kitchen and bathroom.

Some pollutants arrive in our homes via upholstered furniture, such as a new sofa, and clothing, carpeting and paint. If possible, unwrap new purchases and allow them to 'off gas' outside or in a garage for a couple of days before bringing them into the house. Look for low VOC paints and try to stick to natural fabrics for clothing, bedding, towels and soft furnishings; manufactured fabrics release plastic particles during wear and tear.

Keep room temperatures low and ventilate your home now and again by leaving doors and windows open for 5–10 minutes to get the air circulating so that toxins are dispersed before they get a chance to stagnate. Finally, bring nature indoors: researchers at NASA have shown that spider plants, aloes, ferns and other house plants can absorb chemical pollutants[18] and their very presence may also make us feel more tranquil and happy.

How can I reduce all the plastic in my bathroom?

ISSUE

When I look around my bathroom I realize how many plastic tubes and bottles crowd the shelves and line the bath and shower. I like to keep my skin and hair in good condition: what can I do to reduce the number of products that come in plastic packaging?

SOLUTION There is very little robust, independent, scientific evidence that beauty products actually make any difference to the condition of your skin and hair.[19] Your age, lifestyle, general health and wellbeing are much more important, but we are constantly seduced by the promise of beauty, lurching from one product to another, accumulating bottles as we go. So the first task is to review all those bottles and work out which products you actually need.

Don't try to change everything at once. Put all your bottles of products in a big box and take out just those you need as you use them. You may find you only need half of them. As each product runs out replace it with a plastic-free alternative – this way you're not wasting the products you already have and not replacing everything at once, making it gentler on the environment and on your cash flow.

To reduce the number of plastic bottles, buy bars of soap and try shampoo and conditioner bars and mouthwash tablets. If you use a mesh soap saver bag, put the soap in the bag and use it as an exfoliating scrub; or why not try a konjac sponge for cleansing? Made from plant fibre it is reusable, biodegradable and compostable.

You can even make your own natural products for next to nothing. Oat is great for sensitive skin: a cupful of rolled oats in a small cloth bag (or a sock) with a few drops of essential oils attached under the running tap makes a relaxing and soothing bath. Coconut oil from your kitchen cupboard works well for hair and skin smoothing, but use sparingly because coconut plantations can lead to deforestation of rainforests.

For the liquids you can't phase out, save the empty bottles and take them to be refilled: there are more and more bulk refill options around. Search on social media sites for a zero-waste group in your area and you will be able to share the information on local refill options and zero-waste shops.

Finally, try to look for products in glass not plastic containers and if you want to check the green credentials of the brands you are considering and suspect they may just be greenwash, check them at ethicalconsumer.org or with the Giki Badges app.

How can I reduce clothing waste?

ISSUE

Fashion is one of the most polluting and wasteful industries in the world and although it is improving its performance every year it is still failing to reach targets like the Paris Agreement. I can't wear the same things all the time so how can I reduce clothing waste?

SOLUTION Most of the environmental impact of the clothing industry is in production and waste, both of which are fuelled by fast turnover. If we extend the life of our clothes by just nine months it can reduce the carbon, water and waste impacts by 20–30 per cent.[20]

We need to rethink our approach to clothing. We should buy less and enjoy it more. The fashion industry wants us to change our look every year and the cycle of the seasons encourages that. We need to view clothes as long-term commitments not semi-disposable items; reusing winter and summer outfits from year to year. Learning to identify which cut and colours actually suit us means we are less swayed by the new season's shape and style, reducing the number of impulse buys that inevitably don't last in our wardrobes.

As consumers who care about sustainability, circularity is something we can all focus on.[21] One of the ways to extend the life of our clothes is to return them to the marketplace. Clothes trading sites and pre-owned fashion apps are popping up all the time – it's easy to advertise your own old clothes and see what else is on offer. Vintage, charity and thrift shops can be found on every shopping street along with more and more upcycling stores taking in clothes and repurposing them. Have a rummage in your wardrobe – there are bound to be some old favourites that could be given a makeover. If you want to give it a go yourself repurposing clothes is a lot of fun. Making old jeans into draught excluders, jumpers into cushion covers or shirts into bags can be nicely nostalgic. There are lots of ideas available online. Or meet up with friends for a clothes swap – it's a great social occasion and you may find some new treasures and offload some to make some space. Take on the challenge and see if you can go without buying any new clothes for a month, a season or a year.

75% of fashion consumers think sustainability is extremely important.

85% of textiles end up in landfill or incineration when most could be reused.

The fashion industry emits more carbon than international flights and maritime shipping combined.

Are there plastics in my clothes?

ISSUE

I'm doing everything I can to banish unnecessary single-use plastic from my life. I've eliminated plastic bottles in my bathroom and kitchen and now I've started to think about my clothes: are there plastics here too?

SOLUTION The answer is probably yes. Look in your wardrobe and check some labels. The likelihood is that your clothes will contain some synthetic fibres which are all types of plastic – look for elastane (Spandex/Lycra), polyester, nylon, acrylic, polyvinyl chloride (PVC/Vinyl) and polyurethane (PU). Worldwide around 60 per cent of textiles used for clothes contain plastic.[22] If any of your clothes are close-fitting, stretchy, wrinkle-free, sparkly, breathable or waterproof, ultra-light and packable then that is probably down to their plastic content.

Greenhouse gas emissions are inherent in plastics as they are manufactured from crude oil and use fossil fuel energy in production; furthermore they take 500 years or more to biodegrade. And that's not the end of it. When we wear or wash our clothes tiny fibres are released that end up in the oceans and air. On a daily basis we are eating, drinking and breathing in microfibres from synthetic clothes and fabrics.

The good news is that now you know where these plastics are you can avoid them. But, as we just saw, clothing waste is a huge problem so don't just bin everything you have or you're adding to the clothing waste mountain. Wear them to death and make them work for their environmental impact, and when they finally die take them to a responsible recycling centre.

For new purchases of clothes, towels and bedding choose plastic-free, natural fibres: organic cotton, linen, silk, wool and hemp – they may cost more but buying less of what you love will last longer and save you money. To get the performance you have become used to you may have to accept a small percentage of synthetic fibres – even the most sustainable brands contain 5–8 per cent synthetic materials in underwear and sportswear. Have a look at semi-synthetic fibres; tencel, bamboo and modal are all manufactured from plants and are plastic free, but you need to balance other environmental and ethical impacts of chemicals used in production and the manufacturing conditions. All fabrics have environmental impacts so do some research and aim for brands that source the most eco-friendly materials.[23, 24]

Is cotton a good natural alternative to synthetics?

ISSUE

I've worked out which textiles in my house have plastic in them and I want to move to more natural alternatives. But now that I know they all create environmental problems, is cotton a better alternative to synthetic fabrics?

SOLUTION While growing cotton is the livelihood of millions of farmers, the crop depletes soil fertility, demands huge input of water and pesticides, most of it is genetically modified (GM), and its production raises ethical and human rights issues. Organically grown cotton is a better option because the plants are not GM and don't rely on chemicals but because yields are lower they require more land and more water. On the face of it, cotton doesn't sound like a great alternative to synthetics.

Fortunately, changes are afoot. In 2005 the Better Cotton Initiative (BCI) was set up to transform the industry by introducing sustainable practices. It is attracting retailers (most high-street brands are pledging to use only sustainable cotton by 2025) and is supported by the WWF and Pesticides Action Network. By the end of 2018 its standards were met by 19 per cent of global cotton production, a figure that is set to rise. By adopting BCI practices, water and pesticide use is cut by around 40 per cent.

Many ethical brands are already achieving Fairtrade or organic standards for their cotton supplies – these are the ones to support.[25] Check how your favourite brands perform on the ethical consumer website. Look for organic cotton when you buy new or recycled cotton. As an alternative fabric, try hemp: it grows widely round the world and without pesticides, uses only a fifth of the water needed for traditional cotton and a third of organic cotton and it enriches rather than depletes the soil. Or opt for a similarly more environmentally sound choice: linen, which is derived from flax, and as an added benefit it can grow in poor-quality soils.[26]

. . . .

Good to know . . .

• In 2013 the water used to grow India's cotton exports, for example, would have been enough to supply 85% of the country's 1.24 billion people with 100 l. (almost 22 gal) of water every day for a year. Meanwhile, more than 100 million people in India had no access to safe water.[27]

• Diversion of water and pollution of the waterways has severely impacted on the Indus delta in Pakistan, the Aral Sea in Asia and Australia's Murray Darling river system.[28]

How do I reduce the impact of washing my clothes?

Washing my clothes uses energy, water and chemical detergents sold in plastic bottles, and now that I've discovered many of my clothes contain plastic I'm worried about microfibres ending up in the ocean. How can I ensure I am doing my laundry responsibly?

SOLUTION It is shocking to hear statistics that estimate there are now more microplastics in our oceans than stars in our galaxy, but with a bit of thought we can reduce the impact we have when we do our laundry. We've just discussed (page 34) how you can start to eliminate plastics in your clothing and other textiles. Buying the most efficient washing machine when it's time to replace yours will help, and using it as economically as possible will help more (see page 20).

Minimize your number of washes. Clothes that are not worn next to the skin may not need washing as often as you think. Sort through your worn clothes before you throw them in the laundry basket, they probably don't all need washing. Hang them up to air and they may be fine to wear again, while small marks may be sponged off rather than washing the whole item. Delay washing until you have a full load, this reduces friction on the fabric, meaning fewer microfibres are shed, and don't tumble dry – that reduces the life of fabrics and detaches more fibres – whenever you can, hang your clothes out to dry in the fresh air. Both of these choices will save you energy and reduce your carbon footprint as well. You could try to minimize the loss of microfibres by using a specially designed washing bag or a laundry ball that will trap the fibres so that they can be removed rather than entering the waterways.

Buy environmentally friendly products in refillable bottles (or no bottles at all) and avoid synthetic fragrances to reduce toxins in your home. Try natural products that eliminate the need for any chemicals – great for sensitive skin – look out for 'soap nuts' or ceramic and mineral pellets, both of which can be used repeatedly.

· ·

Good to know . . .

• Clothing fibres have been found in fish that inhabit the Mariana Trench in the Pacific Ocean,[29] the deepest trench in the world.

• Studies have estimated that even a small town of 100,000 people would produce around 1 kg (2¼ lb) of plastic clothing fibres each day, all of which enter the oceans.[30]

2.

OUTDOORS

Our first interface with nature is through our gardens, roof spaces, balconies and windowsills. From here we venture out into our local spaces, parks and wilder land. As individuals, our direct influence on nature may seem small on the global scale, but our combined green spaces make a significant difference.

Time outside is fundamental to our physical and mental health. The choices we make outdoors have the potential to connect us more firmly to nature, to protect our environments and to enhance our wellbeing as well as the natural world around us. The greener, wilder and more natural we can make our gardens the more we are helping to halt the dramatic declines in wildlife – particularly in towns and cities.

With just a small amount of effort – and sometimes just by making less effort – there are choices we can make and things we can do that will help insects, native plants and birds to thrive and overall biodiversity to recover.

How can I protect my plants without using chemicals?

I want to keep my plants free from pests and diseases and to help them flourish, but using chemical fertilizers, pesticides and insecticides is expensive and kills wildlife as well as garden 'pests' such as slugs, snails, caterpillars and the rest. What's the answer?

SOLUTION We need to work with nature because those 'pests' are a critical part of the ecosystem on which other creatures depend. Declining populations of pollinating insects, ladybirds, beetles and spiders are having a knock-on effect on birds and small mammals that we welcome in our gardens. There are ways you can ditch the chemicals and encourage a diversity of healthy plants and wildlife.

If you have been using chemicals it may take time for your garden and the soil to rebalance – pests with short lifecycles may recover faster than their natural predators. So you need patience but there are plenty of things you can do to encourage the process while employing other tactics to keep pests under control.

Encourage birds by feeding them in winter and boost their natural food sources by building a log pile or bug hotel for hibernating insects. A small pond will attract frogs to keep down insect numbers, and clumps of nettles bring the nettle aphid, an early season food for ladybirds that will then feast on aphids in the main growing season. Slow worms eat slugs, so lay down corrugated metal for them to hide under. Create gaps in walls or fences or plant a hedge of native species so that these animals can move freely.

Clever and varied planting can minimize the damage caused by pests or act as decoys. Alliums – onions, chives and garlic – repel many pests while planting a few 'sacrificial' lettuce on the edge of your veg patch will attract slugs so you can catch them before they get to the main crop. Slugs and snails are nocturnal feeders so go out with a torch on damp evenings – or scatter cabbage leaves or grapefruit halves to attract slugs away from your plants – and physically remove those that you find.

Mixing up the planting can help reduce outright devastation and avoid recurring infestations. Timing can be important too, for example, sowing later or harvesting earlier to avoid the main pest attack. Barriers such as mesh around soft fruit plants, crushed eggshells and used coffee grounds will deter molluscs, or make mini cloches from plastic pots before they go for recycling to protect young seedlings.

Why can't I use peat in my garden?

ISSUE

What is wrong with using peat-based compost in my garden; after all, it is a natural product but I've read that preserving peat bog habitats is at least as important to reaching net zero carbon as saving the rainforests. Why is this?

SOLUTION The peat used in garden compost or as fuel is formed over millennia in bogs and wetlands. Because it takes so long to form, peat – like fossil fuels – is non-renewable. Ecologically rich peatland habitats occur widely across the globe and their importance goes beyond even their critical role in preserving habitat diversity: in their natural state, peat bogs represent the biggest terrestrial store of carbon: more than all the other vegetation types combined. We cannot protest at the destruction of rainforests or challenge the fashion or aviation industries with a clear conscience unless we are also protecting and restoring the world's peatlands. Lack of awareness of their importance has led to the degradation, with more than 15 per cent of peatland being drained or burned, and when this happens the carbon locked in the ground is released into the atmosphere. Peat bogs are vital defences as the world faces climate change. In its natural state, peat can store up to 20 times its weight in water, reducing the impact of flooding following heavy rainfall and acting as a store for water. Dry, degraded peatlands represent a huge fire risk – peatland fires can smoulder for months, even after prolonged rain and under layers of snow, releasing CO_2 as well as toxic substances.

Peat for our gardens is a luxury we cannot afford. The good news is that you don't need to use any peat; a range of excellent and sustainable alternatives for different soil types and plants – even ericaceous and other peat-loving ones – is available from garden centres. And you can make your own compost: see page 54.

. . . .

Good to know . . .

• In the UK up to 70 per cent of domestic water supplies come from peatland.
• In 2015 fires in Indonesian peat swamp forests emitted, per day, more than the daily emissions from the entire US economy.[31]

• In Borneo, Sumatra and the Malay peninsula destruction of peat swamp forests, principally for conversion to palm oil plantations, has released so much CO_2 that their greenhouse gas emissions are now comparable to those of China and the USA.[32]

What can I do to help pollinators and birds?

ISSUE

Climate change, urbanization, pressure from agriculture, use of agro-chemicals and other pollutants, and invasive non-native species are taking their toll on insect and bird populations. With so many species threatened with extinction, what can I do to help restore biodiversity?

SOLUTION What we do in our gardens, allotments, balconies and window boxes makes all the difference to bees, butterflies, moths and birds – and there are always options to make your space more wildlife friendly. Pollinators need energy to fly; some migrate long distances or hibernate over winter and most of that energy comes from nectar. Grow lots of different nectar-rich plants in blocks for the spring, summer and autumn months, water well and deadhead regularly to help them produce more blooms, hence more nectar. Limit the amount of bare concrete, fences and decking, which offer nothing to wildlife, and plant a mixed native hedge or climbing plants on trellis to provide shelter from rain or frost. If you have window boxes, try growing perennial flowering herbs, such as rosemary and lavender.

Sometimes doing less tidying is all that is needed. Leave wild areas for aphids and larvae (see page 43) plus a pile of leaf mulch that is rarely disturbed. If you have a lawn let one section grow long. You may not go as far as rewilding your patch, but not mowing during spring will benefit wildlife by allowing 'weeds' to bloom.

Plant, insect and bird populations evolve together, so by welcoming or growing native plants you are providing the best food supplies to support a wide range of native pollinators and birds, increasing their diversity. Plants with good seedheads provide winter seed for birds – so resist the urge to cut them off when the flowers die. In months when food is scarce, put out a choice of foods, such as uncooked oats, seeds, fruit or cooked rice; avoid salty food or anything too hard that could choke.

. . . .

Good to know . . .
• Globally the insect population is in danger with more than 40% of species facing extinction over the next few decades.[33]
• Globally, 40% of bird species are also at risk of extinction, though this is decreasing.[34]

• Our gardens are important wildlife reserves, especially in cities.
• The combined surface area of all the gardens in the UK is the same as four of its national parks: Exmoor, Dartmoor, the Lake District and the Norfolk Broads.[35]

How can I help to preserve wildlife in my local area?

ISSUE

It's not only insect and bird species that face extinction: land-use changes including extending built-up areas with more housing, more infrastructure and services impact on habitats and biodiversity. What can I do to reduce that impact and help the wildlife where I live?

SOLUTION It's easy to feel overwhelmed by reports of declining biodiversity but actions that we can take at the local level, however small, can make a difference. If enough people do the same, there's hope that we can reverse current trends – and help ourselves too.

As a start, get involved with conservation work in your area: find out about wildlife trusts, rivers trusts, and nature conservancy and conservation groups. Go along to events and visit local nature reserves and wildlife sites to learn about your area and its wildlife. Lend your support by joining a volunteer force – they always welcome practical help with tasks such as habitat creation and clearing invasive species, or scientific work such as undertaking species surveys or testing water quality, or office-based support, whether responding to calls or being active on social media, providing database support or offering help with fundraising.

Speak up and have your say; many local wildlife groups run campaigns trying to persuade local and national government to make positive change for nature and to save local wildlife areas. You can talk informally to others in your neighbourhood about the issues and try to get them onside – or simply set a great example when you are out and about by taking a bag and gardening gloves with you and picking up litter to prevent it causing harm to wildlife.

Any of these small actions will not only help your local wildlife but help you too by getting you active, perhaps away from your screen and out in green spaces, meeting others and making a difference – positive steps that have all been shown to help mental health and wellbeing – leading to a happier you and a happier planet.[36, 37]

Is native wildlife affected if I grow exotics in my garden?

ISSUE

Invasive non-native species are listed as one of the reasons why our pollinator and bird populations are declining. I have lots of plants in my garden and several of them are exotic species: is this a problem for native plant and animal species?

SOLUTION Almost every country has a rich mix of native and non-native species. Invaders, settlers, traders, explorers and plant collectors have all introduced 'exotics', some of which happily and unobtrusively found a niche. But others become invasive and spread quickly, outcompeting native species and harming wildlife and waterways.

Having exotic species in your garden or pond may be an attractive idea, but the problem is that they typically escape. Seeds and berries transported by birds, insects, wind or water spread plants far and wide – even fragments of plant in the mud on your boots can cause an invasion once they get out into the countryside. Whenever possible, stick to growing native species; they support richer native insect and bird populations, contributing to greater diversity and you needn't worry about them causing a problem for local wildlife and habitats or a legal one for yourself (see fact box).

Garden centres can sell species that are listed as invasive (although they should not have any that are restricted by law); these are not illegal to have in your garden but you should still not release them into the wild. You can find good advice online on which species are non-native and which are native.

14 non-native species identified by the EU pose such a risk that it is an offence to keep, cultivate, breed, transport, sell, exchange or release them (intentionally or not) into the environment anywhere in the EU.

£1.7 billion a year is spent in Britain tackling the problem of invasive non-native species.

60% of non-natives in the UK that harm wildlife and waterways are believed to have come from gardens.[38]

42% of species perceived to be threatened or endangered in the USA are at risk due to invasive species.[39]

If you do have non-natives in your plot, clean off your clothing and footwear before leaving the garden and don't give these plants to neighbours or friends: it just spreads the problem. Or, if you reach the point where you want to get rid of a non-native plant, it needs to be composted on site or go with your garden waste for commercial composting. If you have a big problem with invasive species there may be grants to help you eradicate them responsibly and you can employ specialist companies.

I know trees absorb carbon dioxide; can I plant more?

ISSUE

The top scientists believe that to stay within the safe limit of climate change we need to keep the global temperature rise to +1.5°C (2.7°F)[40] and that planting trees is one of the best options for reducing atmospheric levels of CO_2. Can I help plant more trees?

SOLUTION Worldwide tree-planting programmes could remove two-thirds of emissions from human activities that are in the atmosphere today[41] and governments are making commitments to what seems to be a readily available and affordable solution. Tree planting doesn't require technological advances, everyone can get involved and there are many other positives for the planet: trees reduce flooding and pollution, support a wide range of wildlife and make landscapes more resilient and beautiful.

Whether you want to volunteer to help with planting, you have your own land to plant or you think you can persuade a school, farmer or other landowner to do some planting the best option is to link up to a local tree-planting scheme or woodland trust. Any offer to dig in and help out will be welcome and you will also receive advice from experts on how to choose a tree, where and how to plant it – they can even sell you native species that are right for the conditions in your environment, ensuring them the greatest chance of success.

You can also contact a global organization like One Tree Planted and the Arbor Day Foundation and make a donation to have trees planted globally or you can fundraise or get your hands dirty by doing some actual planting. All the tree-planting schemes have options to sponsor a tree, dedicate a tree or donate a real tree. It's an idea that has caught the public mood, one that can translate concern for the environment into a highly successful campaign online. For example, #TeamTrees made a pledge to plant a tree for every dollar donated. Within two months it had not only raised enough money but achieved (through the Arbor Day Foundation) its target of planting 20 million trees.

You can even grow your own native tree from seed.[42] It will take a couple of years before your seedling is ready to be planted out and until then you can watch its progress and nurture it in your own home.

How can I make my own compost?

I want to reduce my food waste and – now that I know how vital it is to avoid horticultural products that contain any peat – I would like to make my own compost. Where do I start, and are there any tips for what I can and can't use?

SOLUTION By composting you are working with nature (microorganisms) to break down organic matter to create rich organic soil. Start by putting coffee grounds around houseplants or herbs; bury fruit and veg peelings under the soil; or oven-dry eggshells then crush them to a fine grit to sprinkle around plants.

If you have a bit of outdoor space, you can simply put kitchen scraps in a pile and let it decompose over 6–12 months – to speed up the process turn it every few months with a garden fork to aerate the mix but otherwise leave it to its own devices. If your space is restricted you can make compost in a fenced structure or a bin with either a few holes in it or a tap to drain off nutrient-rich liquor (and avoid a stinky sludge forming); whatever the structure the process is the same. Covering with an old piece of carpet or tarpaulin helps to keep it moist but not sodden. (You can also compost indoors; here you will need a bioreactor or worm bin[43] but the basic rules still apply.)

A good mix of organic waste will give the best results. Save up your kitchen scraps and add them with some garden waste; this way works better than making tiny additions but a lot of grass cutting or hedge trimmings can overload the compost, so take some to your municipal composting. You can include: shredded newspaper and unprinted cardboard, bedding and droppings from vegetarian pets (hamsters, rabbits, birds), teabags, coffee grounds, fruit and vegetable scraps, plant and grass cuttings and dead leaves.

Do not compost cooked food, coal ash, meat, fish or bones, cat, dog or human poo, nappies, glossy paper, weeds, woody stems or diseased plants – temperatures in compost are not high enough to kill off seeds and disease pathogens so they will flourish when you use the compost; the other contents attract rats and smell bad.

When you turn your compost watch out for newts, slow worms and hedgehogs, which may be hibernating. Your compost is ready to use when it is dark and crumbly like a good soil. You will soon be reducing your food waste, increasing the organic content and structure of your soil, saving money and the peat bogs (see page 44).

How can I grow my own without a garden?

ISSUE

I'd love to be able to reduce the distance my fruit and veg travels, and to know exactly how it has been produced, but how can I have the pleasure of growing and picking my own fresh ingredients when I don't have a garden?

SOLUTION You don't need a garden to grow your own food. You'd be amazed at what you can grow on a windowsill, balcony or roof space. All you need are a few pots (ideally reuse plastic food pots and tins before recycling or use terracotta or biodegradable containers) or a planter box, a trowel and fork, some peat-free compost (see page 44) and some seeds.

A great place to start is with herbs to boost the flavours in your cooking – basil, oregano, thyme, mint, sage, and parsley all grow well on a sunny windowsill – and they will help your home smell great too.[44] Next try some ingredients – salad leaves and microgreens (the seedlings of edible plants like mustard, beet, sunflower, broccoli and buckwheat) are great for small spaces. Cut off the salad leaves you need, just above the roots, and the plant will keep producing, while microgreens grow quickly and are highly nutritious.[45]

Once you're into the growing mode, if space allows, you can trial using bigger pots and slower-growing plants – how about tomatoes, peppers, green beans and spring onions? Experiment: try different plants to find out which ones work for you and the space you have – don't be put off if you don't have immediate success. Rather than always buying new seed try growing new plants from vegetable scraps – celery, fennel, garlic, ginger, lettuce, bell pepper, strawberries, turmeric and many others will work.[46] Remember to dispose of non-natives responsibly as you do with your vegetable waste.

Be imaginative with the space available – use vertical surfaces to extend your growing area by hanging pots and creating pocket planters. If your sun levels are challenging you can use a fabric shield for too much sun or rig up a clever mirror to direct more light if it is limited.

Once you get the gardening bug you could rent an allotment – it is, after all, just an off-site garden. . . .

How do I avoid wildlife being harmed when I go out?

ISSUE

It seems everywhere we humans go we damage and destroy parts of the natural world. I love being out in the countryside and enjoying wild places but I'm so worried about the impact we are having on the planet I hardly dare go out.

SOLUTION   It's really important for our health and wellbeing[47] to get out into the natural world – and the more time we spend with wildlife the more we understand and appreciate it. Rather than becoming too worried about harming wildlife there are some simple codes of practice you – and your children and pets if they're out with you – can follow.

Every one of us has a responsibility to protect the countryside and the general ethics of treading lightly and leaving no trace – neatly summed up by 'take only photos, leave only footprints' – will cover most eventualities.[48] This means not damaging or removing natural features that we see and appreciate – including plants, rocks and trees – which are all homes to wildlife and form an integral part of a functioning ecosystem. This is demonstrated if we move a rock or log: we immediately realize we are disturbing the home of small animals or insects, so put it back in place gently.

Similarly take home your leftover food and packaging to dispose of it responsibly. Everyone dislikes seeing litter in the countryside, and it can injure or harm wildlife. Avoid naked flames and put cigarettes out carefully – we've all seen the devastating effects fires can have on nature and property – don't risk being the person who starts one.

Stick to paths unless wider access is permitted. This will prevent excessive wear and tear in wild places and disturbance of ground nesting birds or breeding wildlife (particularly if you have a dog with you). If you come across wild animals give them space and cherish the moment; watch and enjoy them quietly from a distance and try not to get between them and their escape route, which might be the sea for a seal or unfenced woodland for a deer.

My local area is covered in litter, what can I do?

ISSUE

Litter is an eyesore, extremely hazardous to wildlife and much that is not biodegradable ends up polluting our waterways. What can we all do to solve this problem and safely dispose of rubbish without relying on hugely expensive clean-up operations?

SOLUTION The most common object found during litter clean-up is fast-food packaging, which attracts hungry animals. They get inside bags, cups, cans, bottles and jars and either become trapped or they choke when they try to eat wrappers and bags. They also get entangled in discarded balloons, can-holders, elastic bands or fishing tackle, which injures or even strangles them as they struggle to get free, while broken glass bottles cause cuts and injuries. So take responsibility for your rubbish by disposing of it properly – ideally by recycling. Keep a bag in the car for rubbish so you are not tempted to throw it out the window. Simple things like knotting plastic bags, deflating balloons and snipping up elastic bands and can-holders can reduce the hazard to wildlife if the waste does end up in the environment.

5,000 calls annually made to the UK's main animal charity (RSPCA) concern wildlife that has been injured by litter.[49]

9 billion tonnes of litter end up in the ocean each year.

An estimated **43%** of Australian seabirds have plastic in their gut.[50]

$11.5 billion is the annual cost of clean-up operations in the USA.[51]

Next, encourage others to act responsibly. Do some litter picking – either on your own or by joining or organizing your own community litter pick. Local councils will often lend equipment to keep you safe. It's satisfying to be taking action and other people may take the hint – a clean area is less likely to attract more litter.

The main reasons people give for littering are lack of easily available bins and being too lazy to find one, so lobby your council for visible recycling and rubbish bins in locations that attract most litter. Report anyone persistently littering and if you pick up litter check where it came from: packaging, cups and bags are often branded. Visit shops and outlets that are a key source of litter and ask them to use biodegradable materials, provide bins for recycling and run education campaigns informing customers of how and where they should dispose of packaging. Waste strategies now put the onus for packaging waste on the companies that produce the packaging, so give it back to them.

How can I ensure my dog's waste isn't a health hazard?

ISSUE

Dog waste is unattractive and messy; it is also an environmental pollutant. Nutrients in their waste disrupt the natural balance, while bacteria, parasites and other disease vectors wash into waterways; posing a danger to wildlife, domestic animals and people.[52]

SOLUTION The answer is clear, and it's often the law: for all urban spaces, popular dog-walking areas, and land used for livestock,[53] collect and bag all dog mess and bin it or take it home and dispose responsibly. The code in wilderness areas away from urban centres or farm animals is more debatable; here a 'stick and flick' approach could be better than more plastic bags. This is a judgement call but it's easy to underestimate the number of dogs sharing the same space on a daily basis and one study suggests the carrying capacity (the load of a pollutant natural ecosystems can cope with) is just two dog poops a day per square mile.[54]

Never be tempted to collect dog waste in a bag then leave it on the ground, on a tree or in the undergrowth – you create a poo bomb wrapped in a highly persistent plastic casing. Even biodegradable bags may be present after three years and compostable ones persist for more than a year.[55]

Ideally the default is: always pick up dog mess in urban areas, on land grazed by livestock or anywhere popular with other dog walkers. In more remote areas, the best option would be to bury all your dog's deposits to minimize the risk to local waterways and wildlife.

There are a lot of dogs in the world. Estimates suggest:

• around **8 million** in England and Wales

• **76 million** in the European Union (excluding the UK)

• **77 million** in the USA.

On average each dog produces **340g (¾lb)** of waste a day.

The bacterial load in **2–3 days' worth of waste from 100 dogs** would be enough to close a swimming beach as a hazard to health.

. . . .

Good to know . . .

• Dog faeces on public and land where animals graze is a serious hazard. Neosporosis, a disease caused by a parasite found in dog faeces, is the most commonly diagnosed cause of abortion in cattle in the UK.

• Near waterways it is vital to clean up. In a study of a 9.4km (6-mile long) stretch of river in the USA, DNA tests confirmed that 42% of the controllable bacteria load in the water was from dogs. The river's catchment had a population of 11,400 dogs.[56]

TRANSPORT

How we choose to travel is perhaps the most important decision we can take to reduce our impact on the planet. We know that resources are finite and air quality is a constant urban concern.

Every time we step outside our homes, we make a choice about how to reach our destination. And this choice has a big impact on how we contribute to climate change and air pollution. The choices are not always easy, the way our travel infrastructure has developed does not always make the transition to low carbon and low air pollution straightforward or even possible. For many short journeys, though, we can make conscious decisions to improve our immediate surroundings; to take polluting vehicles off the streets of our towns and cities and make them safer and more pleasant places for everyone to live in. Many longer journeys are entirely discretionary and at every opportunity we should be questioning our needs and motives behind a journey – cutting back and combining wherever we can. Now could be the time for some tough choices.

Which transport choices will help cut carbon emissions?

Transport accounts for a large percentage of the world's energy use. Every year we travel further and much of the energy powering our journeys still comes from fossil fuels; this means our transport choices play a huge role in tackling climate change.

SOLUTION Every time we plan to make a journey we should go through a thought process, starting with: *Do I need to travel?* Then, if the need is confirmed, work through the options, starting with the lowest carbon emissions: *Can I walk, cycle or use public transport? If not, can I combine trips to reduce my mileage or share travel with other people?*

Transport choices vary depending on the journey you are planning. For local journeys, whenever possible, walk, cycle or use public transport (bus or train). If you have to drive, this chapter has advice on how to reduce your emissions.

For longer, inter-city journeys for which going on foot or bike is unlikely, go by bus, coach or train. If you have to drive, car-share (per head, a full car has near-equivalent emissions to domestic rail travel). Train emissions vary depending on passenger numbers (a packed commuter train emits lower emissions per person than an uncrowded off-peak service) and how it is powered – diesel or electric: if the train is diesel, going by coach is better for lower emissions.

For international travel flying may not be the only option. Can you switch to train or even, if you are going with others, drive? A car with two or more people has lower emissions over quite long distance (say London to Madrid). Train is always a better choice than flying but how much so depends on the fuel being used and the source of the energy. For example, diesel train emissions can be double those of electric ones and the source of the electricity also has a bearing: in France, 75 per cent of the electricity is generated from nuclear power, with almost zero carbon emissions, whereas most of Poland's power comes from coal, with much higher emissions.[57]

In the USA travel is the biggest source of carbon emissions[58] and transport accounts for 34% of an average UK household's carbon footprint. CO_2 emissions per passenger per km travelled are:

- **133g** domestic flight*
- **102g** long-haul flight*
- **171g** average diesel car (1 passenger)
- **104g** bus
- **43g** average diesel car (4 passengers)
- **41g** domestic rail (a mix of diesel and electric)
- **27g** coach
- **6g** international rail (electric, highly efficient).[59]

* excluding secondary high-altitude effects of non-CO_2 emissions

I need to fly: how can I reduce my emissions?

ISSUE

In 2018 the aviation industry accounted for 2 per cent of all carbon emissions, but, while other sectors reduce their emissions, by 2050 aircraft may account for 25 per cent of the global carbon budget.[60] If I have to fly, are there ways to cut emissions?

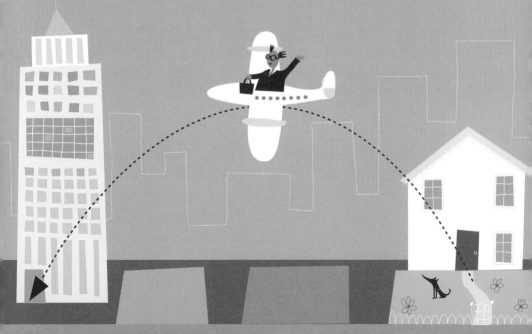

SOLUTION Ultimately, we all need to limit how often we fly and particularly making short-haul flights where there may be other options (see page 67). Nevertheless, for those occasions when only a flight is going to work there are several ways in which you can reduce your impact.

Rather than multi-leg flights try to go direct; take-off uses more fuel than cruising, which is why per kilometre short flights produce more CO_2 emissions (see panel). Try to fly from your nearest airport to minimize the need for transfers and use shuttle buses or trains rather than private taxis and cars. Travel as light as possible – a lighter plane uses less fuel and therefore produces fewer emissions. Choose economy; business and first-class passengers take up more space so emissions per person are three and four times higher respectively (see the panel). Airlines are considering removing business and first class on short flights and some consumer pressure may help.

As a last resort you could consider offsetting the impact of flying by investing in environmental projects around the world. (Carbon offsetting is controversial, though, as it doesn't tackle the deeper structural challenges we ultimately need to resolve. There is scientific uncertainty over the ability of many projects to deliver the claimed carbon savings and reports of inefficiencies estimate just 30 per cent of donated money ever makes it to projects.) Renewables are often seen as the best form of offset – they directly address our reliance on fossil fuels, the central issue causing climate change. Choose gold standard[61] wind and solar projects and offsets through the UN certification website[62] to ensure your carbon and social ethics are sound. Alternatively, support charities that are pushing governments hard to drive climate-change commitments or helping victims of extreme events such as hurricanes and flooding that are becoming more frequent due to climate change.

CO_2 emissions per person per km travelled show the impact of our flying choices:

- **133g** short haul
- **102g** long-haul economy passenger
- **306g** long-haul business-class passenger
- **408g** long-haul first-class passenger.

If we all change to electric vehicles will it actually help?

We know that both petrol and diesel vehicles hugely affect air quality and have a massive carbon footprint. Is the answer to switch to electric vehicles, despite the fact that we still need to generate electricity to power them, and their batteries use rare-earth metals?

SOLUTION   Without doubt we need to end our reliance on gas-guzzling vehicles. Around a third of their carbon emissions comes from manufacturing and the rest from their fuel. In addition exhaust fumes in the form of tiny particles (PM2.5) and nitrous oxides (NOx) directly affect our health. In urban areas these pollutants stay in the air for days or weeks. Diesel vehicles have lower carbon emissions than petrol vehicles but they emit more particulates and are responsible for 15 times the NOx emissions produced by petrol ones; electric cars (EVs) produce no emissions at the point of use.

However, at present, in terms of carbon balance, switching to an EV is not as positive as you would imagine. The power generated from renewable sources is not yet enough to meet all our needs, so an EV powered by a renewables-only energy provider means others must source their electricity needs from fossil fuel energy sources. All engines require rare metals for their components, for example neodymium in motor magnets and lithium, cobalt, nickel and copper in batteries.[63] Mining for these metals impacts on the environment and Tesla predicts there will be a global shortage by 2050.[64] As we move to renewable sources of energy, EV efficiency improves, and fossil fuel cars reach the end of their lives, the carbon balance swings in favour of EVs. Your current choice depends on the type of driving you do and the car you have. Keep your car as long as possible and drive it as little as possible (since a third of the emissions of a car are in its manufacture, it is still better to keep a well-maintained diesel or petrol car until you need to make a change). If you do lots of urban miles in a diesel vehicle, think urgently about switching to electric to cut air pollution. Finally, if you do get a new car – buy a small one and share it more.

Globally, **3.8 million** premature deaths are caused by exposure to air pollution.[65]

In the UK vehicle emissions cause **8,900 premature deaths**, five times the number killed in traffic accidents.[66]

An urban mile (1.6km) of congested diesel vehicles takes about **12 minutes** of life from the rest of the population, compared to **3.5 minutes** for petrol vehicles and **30 seconds** for electric ones.[67]

Will it make a difference if I leave the car and take a bike?

Cars cause so many environmental problems, particularly in our cities, but they too often seem the easy option: would cycling really make any difference?

ISSUE

SOLUTION There are environmental, health and economic benefits to be had when we leave the car and get on a bike. Conventional push bikes are about a tenth of the weight of even the lightest of cars and therefore use a fraction of the energy and raw materials in manufacture, servicing and disposal. Riding a bike uses only the energy you consume in burning calories, saving on fossil fuel use and emissions. Cycling 10 kilometres (6 miles) to and from work each day rather than commuting by car could save 1,500 kilograms (1.47 tonnes) of greenhouse gas emissions a year.[68]

Because a bike has zero particulate or nitrous oxide emissions, cycling also reduces your impact on air quality significantly. By leaving your car behind you will help to cut congestion and keep traffic flowing, reducing your own impact and that of others too.

Using your bike instead of the car can be life-changing for your health and wellbeing. We have all become so sedentary and even short journeys, say a daily cycle of 15 minutes each way, will achieve WHO recommended exercise targets; factor in the reduced stress of not having to pay or find parking, the sense of freedom and connection with nature and you are winning already.

Financially you will winning too – transport is the second biggest household expenditure behind food. A bike is the equivalent of 1 per cent of the costs of buying and maintaining a car. Not making that daily 10-kilometre (6-mile) commute by car will save you a substantial amount in running costs and depreciation.

Another consideration is an electric bike – they offer a low-carbon option for many journeys that are not practical on a conventional bike. E-bikes are much more efficient even than a small electric car, you can travel 20 times further on the same amount of energy.[69] But don't dispense with your push bike just yet – e-bikes take more resource and energy to produce; they, like EVs, use batteries and magnets containing rare metals and you lose out in the financial and health and wellbeing stakes too.

Sometimes I need to use my car: how do I cut emissions?

ISSUE

I am leaving the car at home as much as possible by walking, cycling and using public transport; but sometimes the car is the only solution to getting around and achieving certain tasks. Are there things I can do to reduce my emissions when I do drive?

SOLUTION There are many small changes, including good maintenance and driving habits, that can have a significant impact on fuel economy, hence emissions. A well-maintained vehicle saves on fuel and emissions and makes it safer.

First, think about the car. When you come to replace it, consider buying the smallest, most fuel-efficient vehicle you can for your needs. Look at the mile per gallon (mpg) equivalents – even EVs vary in their carbon efficiency. Don't buy a four-wheel drive unless you really need one and ask yourself how often you actually need the extra storage – could a back-mounted rack or cargo rack give you extra capacity when it's needed? These are a better option than roof boxes and roof racks (see below) – cheaper, more aerodynamic and therefore much more fuel efficient. Remember to take them off when not in use. Next, think about how you drive. Drive a little slower and avoid hard acceleration and hard braking. Take the shortest route available and combine trips to cut out excess mileage; a cold engine is also less fuel efficient than a warm one, so better to do all your errands in one trip rather than lots of short outings. Don't sit in the car with the engine idling, it only takes around 10 seconds of fuel consumption to restart the engine, so if you are stopping for longer than that, turn it off. Finally, think about what you take with you and reduce the amount of unnecessary luggage you carry around on a daily basis: the lighter the car the less fuel it is using. Conversely can you take others with you to save someone else's journey and can you offer your journey to a car-share scheme?

. . . .

Good to know . . .

• Under-inflated tyres lower fuel mileage by about 0.2% for every 1 psi drop. The right oil, keeping the engine tuned and running smoothly can increase fuel efficiency by 4%. Repairing a serious fault can increase efficiency by 40%.[70]

• Accelerating and braking hard can lower fuel mileage by 15–30% on the open road and by 10–40% in stop–start traffic.[71]

• A large roof box reduces fuel economy by up to 17% on the open road, whereas a back box only reduces economy by up to 5%.[72]

4.

ON HOLIDAY

Taking a break is precious time. Our lives are so busy we really long to set aside normal activities, to get away from home, leave behind the strains and stresses of our daily lives, and to dial down our anxieties and worries.

A holiday can be a time to unwind or an opportunity to reconnect with family and friends, to expand our horizons or experience amazing cultures, wildlife and landscapes. It is an opportunity to see and enjoy the world beyond our everyday lives.

This downtime, however, cannot be an excuse to set aside all responsibilities. In place of exploitation and consumption, let's get something more from our holidays, let's all be responsible travellers – expanding our minds and our horizons but thinking about our destinations and the impact we may be having when we get there.

At home or abroad, we should all tread lightly on the world and leave it in a better condition for the generations to come.

How can I be an eco-conscious tourist?

ISSUE

I want to go on holiday as a break from the stresses and strains of everyday life but I still want to be a responsible traveller. It can be easy to disconnect too much; how do I ensure that my holiday does not have a big environmental cost?

SOLUTION Historically, tourism has been exploitative – using and abusing the natural capital and cultural resources of top destinations. We still want to visit new places and see the extraordinary diversity of natural habitats, wildlife and cultures. We can help to protect these precious environments and support local communities who rely on tourism so that they too can protect their natural and wildlife resources.

Think about when and where you are travelling to in a bid to minimize the impact of over-tourism (see page 82). Go with the intention of blending in, experiencing different parts of the world as the local communities do. Enjoy sampling regional produce that is in season, drink locally brewed beers and soft drinks rather than familiar imported brands. When you travel use public transport. This approach will give you new experiences, more understanding of the community and culture, plus you will be supporting the local economy and keeping both carbon emissions and costs down.

The waste generated by an influx of tourists can be overwhelming for a local community so be conscious of the waste you generate, minimize it as you would at home by taking reusable bags and water bottles and try to find out about and abide by the local recycling and waste management systems.

If you visit wildlife experiences and 'sanctuaries', look for evidence of habitat protection and animal-welfare credentials. The most impactful way to be an eco-influencer is to post online reviews – businesses rely heavily on review sites to sell their products and once providers understand that their customers are eco-aware they will respond by being more eco-conscious. Use the same level of scrutiny and apply the same principles of environmental and social ethics that you would at home. Questions the practices you see, within the context of a different location and, possibly, culture to ensure that your visit is as sustainable as you would hope. The sustainable tourism industry is growing fast so if you need help investigating destinations, tours or accommodation then search out responsible travel companies that you can trust to source environmentally and socially responsible holidays.

How should I pack
when I travel?

ISSUE

Travelling and going on holiday are great experiences, opening our eyes and minds to the amazing cultural and environmental diversity in our world but sometimes our bags just seem huge and full of plastic waste. How can I travel lighter?

SOLUTION When you pack to travel the standard environmental mantra of 'less is more' still applies. Whatever your mode of transport you will use less energy and keep your carbon emissions to a minimum the lighter you can pack. It also supports the local economy and adds to the experience if you buy what you need at your destination rather than taking everything with you.

There are some simple ways you can reduce your plastic and waste burden on your destination too. Separating clothes, shoes and other items into cloth or dry bags when you pack keeps your bag organized and makes it easier to find everything. Once you arrive at your destination and unpack you have a supply of reusable bags for souvenir shopping, trips to the beach or for keeping food and spare clothes separate in your day pack as you hike or explore – no need to pick up any plastic bags.

Solid shampoo, conditioner and body soap bars help reduce the need for plastic bottles and generally take up less space. Pack a couple of beeswax wraps as a plastic-free alternative to boxes and bags when you want to take snacks for the day.

We often worry about the quality of water on holiday and buy bottled water – an average family of four can get through 150 bottles in a two-week holiday and are less likely to recycle as rigorously as they would at home. To reduce your plastic use take a reusable water bottle, then you can buy in bulk and decant rather than buying small bottles. Even better, take a refillable filtration bottle so you can go plastic-free – they can eliminate over 99.9 per cent of microbial contaminants including viruses, bacteria, chemicals and heavy metals from freshwater sources. And don't forget to pack your reusable hot drink cup or flask so you don't need to use disposable cups.

Take a quick-dry towel – less bulky than a beach towel and you can slip one into your day bag for use as a picnic blanket, scarf, beach wrap or sun shelter. As the name suggests these towels dry quickly so you don't end up with a soggy bag and it saves your hotel or accommodation having to wash extra towels.

How can I holiday and not contribute to over-tourism?

ISSUE

Local residents are turning against tourism in some popular European cities, and Thailand and the Philippines have resorted to banning tourists from some islands; how can I visit amazing places without contributing to over-tourism?

SOLUTION Many picturesque cities and islands are suffering from very large numbers of tourists. Around 105 destinations have been identified by an EU report[73] as being under pressure; from Paris and Amsterdam to Angkor Wat and Easter Island, the problem is worldwide. The issue is that tourism is poorly distributed – half of all tourists visit the top ten destinations and every year more people visit the tiny remote Easter Island than go to the whole of Bangladesh.

The sheer numbers of tourists have rocketed with the expansion globally of a middle class, students taking a gap year, retired people wishing to see the world, cheap flights and cruises and digital connectivity whetting our appetites and feeding our curiosity. And many millennials, it appears, prefer to spend their money on experiences rather than possessions.

The volume of visitors to key sites creates five main problems: locals become alienated, the tourist experience is degraded, local infrastructure is overloaded, there is damage to the natural resources and there are threats to the local culture and heritage.

When you are planning a holiday, think about going to lesser-known sites and destinations – take the road less travelled. Don't just follow the media and social media-driven features of modern tourism, be more imaginative and you will be rewarded with unique experiences. Very often a little bit of extra research will reveal that the major attractions are replicated just a few miles down the road with very similar features and cultural sites.

If you do go to the most popular destinations, try to time your visit to be outside of peak season. This is good for the destination and good for you too – prices usually come down and you will get a warmer welcome from the locals without having to push through crowds of tourists with selfie sticks poking you in the eye.

Are there sunscreens that don't harm aquatic life?

ISSUE

Many products manufactured to protect our skin from the sun contain harmful chemicals that are entering rivers and seas. Coral reefs, already under threat from rising sea temperatures, over exploitation and pollution, are particularly vulnerable. So how can I protect myself?

SOLUTION Overexposure to the sun can cause skin cancer but protecting yourself from harmful rays does not need to be at the expense of the environment. Sunscreens containing damaging chemicals are being banned in the top tourist destinations of Hawaii, Florida Key West and the Puerto Princesa Subterranean River National Park in the Philippines (one of the New Seven Wonders of Nature). They prohibit products containing oxybenzone and such bans are likely to spread. You can't scrimp on protection from the sun but you can reduce the amount of sunscreen you use and still get the vital skin protection you need if you wear a rash vest or long-sleeved swimming costume when you swim, cover up on the beach, wear a sunhat and stay in the shade.

If you do use sunscreen, to reduce direct losses into the environment opt for a cream rather than a spray to keep it on your skin, not on the surroundings. Try to use products that are free from known reef-damaging chemicals (see the panel) or bearing the Protect Land + Sea (PL+S) Certification seal. Avoid sunscreens claiming they are 'reef safe', which is an unregulated term, and be aware that ones that are biodegradable can still be reef damaging.

Whenever you buy sunscreen ask for products that have the Protect Land + Sea Certification seal or ones that don't contain the ingredients listed here. Use your consumer power to drive change and remember that fake body tan lotions and other SPF products may contain these chemicals, so try to avoid them by checking the label for ingredients on all your skin care products. Whatever you decide, you do need protection from the sun – skin cancer is a very real health risk.

Each year an estimated **14,000 tons** of sunscreen washes off when we swim or shower, ending up in coral reefs.[74] Chemicals are listed by HEL, the Haereticus Environmental Laboratory, in the USA as known pollutants:

- microplastic spheres/beads
- nanoparticles such as zinc oxide or titanium dioxide
- oxybenzone
- octinoxate
- 4-methylbenzylidene camphor
- octocrylene
- para-aminobenzoic acid (PABA)
- methyl, ethyl, propyl, butyl or benzyl paraben
- triclosan (TCS).[75]

How can I influence others to be eco-conscious tourists?

Tourism impacts issues like climate change, water scarcity, poor animal welfare and damage to habitats and biodiversity. We need to raise the number of tourists who take responsibility for these impacts and are willing to make choices that reduce or improve the situation.

SOLUTION   To effect real change, it is no longer about simply acting responsibly ourselves. There are choices we can make that help to support an environmentally sustainable travel industry, and the places we value as tourists, as well as the entire planet. We must now commit to sharing that knowledge with friends, family and neighbours about how they too can make responsible travel choices. The next stage for growth lies with us, as consumers. If more of us seek change and expect sustainable choices we increase demand and the industry will supply it. Why not pledge to tell your family and friends about the steps they can take to become eco-conscious tourists?

Talk to people, use social media and post reviews on tourism and travel review and advice sites. Businesses rely heavily on these sites to sell their products, so use your voice to identify their environmental credentials. Shout about organizations and companies that provided you with responsible travel options. Help others to see what it means to be an eco-conscious tourist and how you managed to make that choice. As you start to talk about eco-tourism more you will find others with a similar interest in it. Share resources and expand your combined knowledge. Look for travel awards that highlight the industry's best responsible and sustainable operators to help you select and vote for the ones that deliver. Reward those environmentally responsible operators by using them and showcase them positively to your friends and family; make it clear why you made the choices you did and why it was worth it.

Tell everyone about your travel hacks to reduce plastics and waste (see page 80), how you make transport choices, thinking carefully about the places you stay and where you choose to eat. Spread the word and help make every tourist eco-aware.

. . . .

Good to know . . .

• In 2019, for the seventh year running, global tourism exports grew faster than the global export of goods. [76] Tourism consistently outpaced the global economy until borders were closed due to the COVID-19 pandemic.

Which type of accommodation is best for the environment?

ISSUE

I found a study that shows that homestays are more environmentally friendly that hotels. Then I learned the study was carried out on behalf of an accommodation-sharing site, so can I believe the findings? Are homestays and B&Bs a more environmentally friendly option?

SOLUTION It is not possible to answer this question definitively. The main conclusion from the study[77] was that visitors who choose homestays are more likely to use public transport, local services and recycle, which says more about people opting for homestays than the service they received.

To be confident that your accommodation meets your criteria for environmental friendliness do your research or book with a travel operator that you can trust has done the research for you. What you need to look out for in your accommodation is indications that the providers are thinking about their supply chains and trying to reduce waste. Check for locally sourced, ethically produced, short supply chains in their food, bedding and beauty products; and you want to see that they are reducing water use and looking for circularity in their food and other waste. Find out if they are reducing their carbon footprint and helping you to make your visit low carbon. They should give information on how to get to and from the accommodation by public transport, have ideas for cycle or walking tours from the door or even bus tours to reduce reliance on cars and taxis. They may even provide bikes for you to borrow.

Look for the following: is careful energy use encouraged with, for example, thermostats in the rooms, TVs you can turn off, notes to remind you to unplug appliances and switch off the lights when you go out. Does their restaurant have a good variety of plant-based meals that match the quality of the meat-based dishes?

It is worth remembering that hotels are regulated while homestays are not. Regulation means local planners ensure that utilities, transport hubs and local communities are not overloaded by the additional residents. Look for clues that the property on the accommodation-sharing site really is a private home; 'commercial' homestays in very popular destinations are squeezing out local communities.

By asking questions of our accommodation providers we can actually drive more environmentally friendly choices as the market responds to what we ask for; and that could be a homestay, B&B or hotel.

How can I minimize my impact at my accommodation?

ISSUE

I strive to be environmentally conscious at home and consider the best options for travelling but in terms of expectations for services such as food, drink, use of water and energy at my accommodation, how can I ensure I am not overstretching the local resources?

SOLUTION Having considered environmentally friendly accommodation and what you may demand from it, now have a think about how your demands and behaviour can drive and impact on the local and wider environment. For example, how close is your accommodation to the sights you want to see during your stay? Book somewhere near where you want to spend your time to minimize carbon-hungry travel at your destination but also think about spreading the load to avoid over-tourism (see page 82).

Think carefully about water conservation. Many holiday destinations are in water-vulnerable areas – so take short showers rather than baths and turn off the tap when brushing your teeth. Comply with bedlinen and towels policies – don't ask for them to be changed within a week unless absolutely necessary and hang towels up – the usual policy is to wash anything left on the bathroom floor.

Think about energy efficiency – turn off appliances when they are not in use and don't leave your phone and devices charging overnight or plugged in when they are fully charged. Turn the thermostat down or the air-conditioning up by a degree to two; there's no need to stick at a steady 20°C (68°F) – we're still comfortable between 18 and 24°C (64–75°F). Be careful not to pump up the heat or AC and then leave doors or windows open and remember to turn it off when you are out for the day. Resist the 'we're paying for it' mentality!

Be mindful of seasonality and place. Don't demand orange juice or other fresh produce out of season – there are bound to be locally produced goods you can enjoy, and you will be developing an understanding of local resources. You want your accommodation to provide an environmentally friendly service within the constraints of its location so don't expect or ask for western decadence in a developing country or capital city provision in a remote location. Hosts will often bend over backwards to provide what their guests request but at great cost to themselves, and the planet.

5.

AT WORK

A significant part of our lives is spent working and there are choices you can make to ensure your work environment – and time you spend there – is as green as possible.

How we travel to and from work, how we behave at work, feed ourselves at work and the influence we have on our colleagues and management at our workplaces are all opportunities to be an eco-hero. We may feel like small cogs in a big machine but even small choices and incidental conversations can lead to

significant changes when we influence a bigger group or alter
a policy for an entire workforce.

There are real opportunities to make real impacts on reducing
our use of plastic, the amount of paper we waste, or the type of
food and drink we consume, as well as ways of reducing emissions
and saving energy in our offices, schools and colleges. Just
starting the conversation can be the beginning of something big.
Don't think change is beyond your influence – give it a go.

We use so much paper, how can we reduce it?

ISSUE

Paper use continues to rise, contributing to carbon emissions, water consumption and deforestation. Tree plantations for paper pulp replace biodiverse mixed woodland, depleting the natural value of the land as well as its carbon-absorbing capability. What can be done?

SOLUTION Recycling rates for paper are already high so the priority must be to reduce use. A paperless culture is helped by hot-desking – it's harder to accumulate paper if you're at a different station every day. IT solutions make it easier: billing and banking can be done online; notebook software makes for flexible note-taking and you can append to agendas and presentations and share information after meetings – no need to type up handwritten notes. Handouts, presentations, lesson plans and so forth usually end up in the bin, so send them digitally or display on a screen for everyone to see and display signs to 'think before you print'.

When you must print, use recycled thin (60gsm) paper for everyday – save the heavier weight for formal documents. Set margins to a minimum and use the smallest user-friendly fonts. Set double sided as the default and think carefully about how many copies you need. For information materials check you are not contributing to junk mail. Keep communications strategic and targeted – only give leaflets and brochures to people on request rather than to everyone.

Many print systems require you to log in before you print, eliminating unattended printers shooting out unnecessary duplicates and errors, they also allow you track how many copies individuals make. Seeing your personal paper use can motivate you to reduce your printing. Take the initiative: encourage a paperless culture at your college or office then close the loop on paper use. This means knowing exactly where all your waste paper ends up and ensuring it is going to recycling not landfill.

It takes **10 l.** (2 gal) of water to make one A4 sheet of paper and the average office worker uses **10,000 sheets** of paper every year.[78]

8.8 million tonnes of paper recycled in 2008 prevented **11 million tonnes** of CO_2 equivalent emissions.[78]

Globally, paper is one of the most recycled waste streams – recovery rates range from **70 to 90%** for the UK,[78] USA,[79] Europe[80] and Australia.[81]

Despite good recycling and recovery rates, more than **100 million** trees are pulped annually to produce junk mail.

Our workspace is always too hot or too cold: what can I do?

ISSUE

The temperature we work at has a huge impact on energy consumption and our productivity. The trouble is, the 'ideal' temperature is not the same for everybody – I may be too hot while other people are too cold, and some are just right. What's the best option?

SOLUTION No one can agree on the right temperature. It is quite common to be working in a room or office where one person has a desk fan running and another is wearing their coat or a shawl. Meanwhile we are using huge quantities of energy heating or cooling our buildings.

Try to get your co-workers to accept a set temperature in winter and summer, while also suggesting that you adopt a slightly cooler office temperature in winter and a warmer one in summer. At least you will know what temperature to expect when you arrive; most of us dress for the seasons for the journey to work and are prepared for seasonal variation. You can save a lot of money with one or two degrees and your carbon footprint will be much improved.

Working at the wrong temperature affects productivity. The way to combat the temperature battles at work is to adjust your own behaviour to control your body temperature so that you stay comfortable and productive. Dress so that you can add or shed layers as you need to. Wearing natural fabrics is better for the environment and allows your body to breathe. Keep hydrated so that your personal climate control and brain are working most effectively – if it is chilly opt for a nice warming drink – getting up to make the drink will keep the blood flowing and help keep you warm too. If you are getting too hot, opt for a cool drink for hydration, quickly run your wrists under cool water and take frozen fruit snacks to keep in the fridge.

The top energy-saving behaviours in the UK give the following savings over a year in terawatt hours (TWh):

• turn thermostat down by two degrees from 20°C (68°F) to18°C (64°F) **(33TWh)**

• turn thermostat down by one degree from 19°C (66°F) to 18°C (64°F) **(16TWh)**

• delay the onset of heating from October to November **(11TWh)**

• wear a thick jumper indoors in the heating season **(6TWh)**[82]

However, it's hard to find the ideal temperature: a study by Cornell University showed employees made 44% more typing errors when office temperatures were low (20°C/68°F) than when they were warm (25°C/77°F).[83]

Are plastic water dispensers and bottles avoidable?

Our culture is one of eating and drinking on-the-go, whether at work, on the daily commute or in our lunch breaks. We know how important it is to stay hydrated, so how do we achieve that without using plastic dispensers and bottles?

SOLUTION It's not rocket science to cut out single-use plastic bottles, but we need to encourage institutions to change. Nearly every office and school in the western world has fresh water available on tap. Water fountains and kitchens that offer a refill service should be standard at almost all locations, eliminating the need for plastic water coolers and single-use plastic cups and bottles.

We also need to alter our behaviour. As part of an awareness campaign carried out at a large office in London, staff were provided with reusable metal water bottles to gauge the effect on the use of plastic bottles both in and out of work over an extended period. The result was that staff cut their plastic consumption by over 50 per cent – and enjoyed the experience.[84] As a bonus, 63 per cent of participants ended up drinking more water: plastics were reduced, helping the planet; participants were more hydrated and felt good about what they were doing for the environment – improving their health and wellbeing; and the business benefited because the staff had better mental focus and productivity. It's a win-win-win. These kinds of experiments can help to show that providing reusable bottles and a bit of awareness-raising works, and it can save money too.

Anywhere you see institutions such as hospitals, libraries or businesses and schools with plastic bottles or canisters of water ask their managers why they are not providing a water fountain or tap where people can refill their bottles. Encouraging sustainable lifestyles is an important part of any responsible organization's programme and your workplace or college should have a plastics policy showing how they are going to be reducing their single-use plastics – give them the evidence and encourage all organizations to get involved in the movement.

Total body water is around **60%** of our body mass and deficits of **2%** of body mass through dehydration can affect aerobic performance and cognitive function.[85]

More than **90%** of the plastic in office bins is single-use food and drinks packaging that people have purchased from shops.

How do I lower the impact of commuting to work or school?

In many countries policies on school choice mean that children no longer go to the school closest to where they live; similarly the way we live now means we often can't live close to our workplace. How can we minimize the environmental impact of our need to commute?

SOLUTION We need to ask some more questions. Is it really too far to walk or cycle? A half hour walk at each end of the day achieves WHO exercise targets, is a way to socialize with friends or colleagues and makes us more relaxed, enthusiastic and productive. The same goes for cycling: 12 minutes of aerobic exercise improves the attention of school children.[86] See if your school or workplace can make it more convenient with accessible, secure cycle parks and places to store clothing and helmets. Companies can encourage cycling with cycle-to-work schemes and shower facilities.

The move to public transport can be expensive. Companies may help spread the cost with season ticket loans, and local authorities often provide reduced charges or free bus or train travel for school journeys: find out what is on offer.

Do you have to go in? The answer is probably yes if you are a student but many employees opt to work from home part-time. Counterintuitively, this can fuel longer commutes as people who go to the office only two or three days a week will tolerate a longer commute and live further from work.

When do you need to travel? Flexible working hours to avoid peak times spreads the load, reduces emissions and air-quality problems as well as time spent in traffic jams.

Finally, if you are driving, can you share the trip? There are car-sharing schemes in many companies that will help you find someone making the same journey. This shares the cost of travel, the burden of driving and reduces emissions. Another win-win-win.

The global average daily commute is **40 minutes** or **12 days** a year for a full-time worker.[87]

In the UK it is nearly an hour and in London up to **74 minutes**.[87]

In the USA the average, one-way commute is **26.1 minutes**, varying across the states from around **45 minutes** in Washington State to **30 minutes** in California and just over **15 minutes** in South Dakota.[88]

In Australia **7 million** people drive to work every day; only **450,000** cycle or walk, a pattern repeated in all OECD member countries.[89] Traffic delays and congestion at peak times in Australia's six major cities account for around 13 million tonnes of greenhouse gas emissions each year.

How can I help raise environmental awareness?

Although large numbers of people now engage in and are concerned about environmental issues, plenty more don't realize or believe that it's down to all of us to take action or are unaware of the implications of inaction. How can we raise awareness of these crucial issues?

SOLUTION We need to keep talking about environmental issues. Educational institutions, workplaces, youth centres and churches provide opportunities to reach a large group of people at once. Positive actions you can take include holding meetings or showing films in a lunch hour or at the end of the day. The 'Blue Planet effect' works – eco-champions Sir David Attenborough and Greta Thunberg are inspirational: use their words and images to help explain the messages – there are many free resources online.[90, 91] Alternatively, invite a local expert (someone from a university, wildlife organization or charity, or an NGO who really understands the problems) to explain the issues to you and your peers. Or organize a campaign (see page 99)[92] to get people to think about their environmental impact: have plastic-free days or use financial incentives to ditch single-use plastic or make reusable bottles, cups and glasses compulsory. You could encourage people to think about what they eat by persuading your canteen to serve only plant-based food on some days or encourage friends and colleagues to get outdoors to appreciate what we need to save.

Environmental concern is high[93,94] but statistics from the UK show a mixed reaction:

• **74%** of people are concerned about climate change

• **70%** of 17–18-year olds are more worried about climate change than they were a year ago

• **77%** of people think climate change should be part of the school curriculum.

At the same time, more than **a quarter** of people think the threat of climate change is overexaggerated and just **under a quarter** are not concerned about climate change at all.

. . . .

Good to know . . .

Environmental action is mixed:

• 95% of the UK population recycle their household rubbish and a similar proportion use their own shopping bags.

• Less than 10% have reduced their meat consumption for environmental reasons.

• Only 35% would accept higher taxes to tackle climate change, compared to 62% in favour of increased taxes for health spending.

What eco-friendly activities can I do with my friends?

When I go out with my friends we have fun but it seems that sometimes we could be contributing something more to benefit the environment. Are there activities that we could get involved in that are cheap, that can help the planet and still be fun?

SOLUTION There are many eco-friendly activities you can do at home with friends like clothes swaps, upcycling old clothes and furniture, making beeswax wraps to replace plastic food bags and growing plants and seeds (see page 56).

If you fancy more of a day out there are many environmental organizations that run volunteering activities and there will be groups operating wherever you live that would welcome some contribution from you; look around your local area and investigate hospital gardens, community wildlife centres, and wildlife, woodland and rivers trusts. Get in touch and ask about volunteer activities and how to get involved.

One thing you might like to consider that can be a lot of fun and contributes to our environmental knowledge is getting involved in citizen science. Citizen science is where members of the public or amateur scientists collect data or carry out research, usually led by or in collaboration with professional research scientists. Involving ordinary citizens vastly extends the research platform that would otherwise be impossible or impractical for scientists to cover and therefore improves the validity and value of data sets and has great potential to advance science, influence policy and change our view of the world. Online organizations at both the national and international level have web portals that have resulted in some highly successful citizen science research projects.

Among the most popular environment action websites that include hundreds of projects to whet your appetite are iNaturalist,[95] eBird[96] and Zooniverse.[97] There are a multitude of active projects at any one time, including hands-on environmental ones that you can take part in while outdoors, looking out of the window or online. Because the projects combine the observations and contributions of volunteers worldwide, discoveries can be made and results obtained on a scale and at a speed far greater than funded research by professional scientists could achieve alone. It's a brilliant example of how fruitful online collaboration by ordinary people can be and brings you a real sense of active involvement in adding to our understanding of the world.

What is the circular economy and how can I introduce it?

We talk about waste in the system, our 'throwaway society', and the problems of declining resources. I'm also hearing people mentioning the circular economy: what does that actually mean and how can I persuade the management where I study or work to get involved?

SOLUTION In the post-war era, right across the western world, the standard economic model became a highly linear system in which we take raw materials, make something from them, use it, throw it away and start again. The circular economy means we keep materials and products in use, constantly feeding materials back into the system, so that 'waste' from one process becomes the raw material for another, meaning that we are not constantly depleting resources, and the need for landfill sites and incinerators is effectively designed out – something generations before the Second World War would probably recognize. At present only around 9 per cent of our global economy is circular.[98]

We can employ this philosophy where we study or work. If the lifecycle of products and materials is considered from the outset we can retain resources' maximum value for as long as possible. You may be able to see immediate opportunities, for example in reducing and recycling food, paper and packaging or refurbishing and reusing IT equipment or furniture. You could ask your college or workplace to look beyond their own organization and consider their suppliers. Strategic procurement has the power to shape the environmental impact of supply chains. Do they monitor and report on their environmental impact? We should demand circularity from organizations, so that they minimize transportation, use recycled materials in packaging and product design, and manufacture products that can easily be disassembled and recycled.

Start a conversation to encourage colleagues to adopt a more circular economy. You need to find a route to someone in a position to make decisions and to persuade them using your enthusiasm for creating a better world. There are toolkits that can help you with this.[99] If they can't immediately see the benefits, remind them that a reported three-quarters of millennial consumers and employees consider environmental sustainability to be a priority[100] and will reward them with their custom and by working for them if they can demonstrate that they are taking their environmental impact seriously.

6.

FOOD & SHOPPING

With every transaction we make we are driving consumption and supply chains that have an impact on the ecosystems that make up our complex and dynamic natural world. Reckless consumption has led to overexploitation of natural resources. Our global markets and year-round supply of food mean even the most mundane of products are often shipped vast distances, exploiting natural resources on the other side of the globe and emitting greenhouse gases in transportation, and contributing to biodiversity declines and climate change.

We need to make our systems work *with* nature – reducing the use of resources, improving animal welfare and ensuring environmentally sustainable management of the land. We need to be conscious and strategic consumers, to think before we buy.

Making informed choices can change the impact of our shopping habits, influencing the availability of goods into the future and securing humanity's place in the world. We owe it to ourselves and our planet to choose products that promote a better world.

How can I ensure I'm doing all I can to reduce food waste?

We often blame our supermarkets for food waste but the amount they waste is just 0.5 per cent, whereas the average European household throws away 25 per cent of its food. What can I do to minimize food waste at home?

SOLUTION There's a lot that we can do to stop food going to waste. First, and most importantly, have a plan. Work out a meal plan for the week; check your cupboards and write a list of exactly what you need before you go shopping. Buy loose fruit and veg so you only buy what you need and don't be tempted by bulk-buy offers unless you know you will eat the extra or have someone to give it to (or put it in the food bank). These tactics will help cut both food and packaging waste and will mean you don't have tempting titbits lying around.

Make sure you bring items with short sell-by dates to the front of the fridge and cupboards so they don't get forgotten and left to fester. And don't throw produce out just because it has reached the best-before date (or eat it blindly if it hasn't) – always give it the sense check – if food looks, smells, feels and tastes right then it is probably alright, whatever the date on the packet.

Try to eat everything you buy – be innovative in your cooking to use up every last bit of food. Many bits of fruit and veg that we discard can be used, even the tough woody bits can go into casseroles or soups or at the very least be used to make stock. There are apps and websites to help with inspiration.[101]

Making sure your fridge is set to lower than 7°C (45°F) and using your freezer can save a lot of unnecessary food waste; cook in bulk and freeze for another day, excess fresh produce – provided the water content is not too high (for instance, cucumber, melon and lettuce don't freeze well) can be frozen, along with leftovers from meals.

If you have excess usable food, then look for ways to share it in your community. There are food-sharing apps for distributing food locally and friends and neighbours often welcome bits of food. Share the love and save on food waste and money.

Which eggs have least impact on the environment?

Eggs are nutritious, cheap, versatile, good for weight-loss and high in antioxidants. They even come in their own compostable packaging. But, as we are finding with everything, there is an environmental impact: how can I choose and use eggs to minimize their impact?

SOLUTION The main environmental impacts associated with egg production are greenhouse gas emissions and soil and water pollution, and these derive largely from feed production and manure management. The type of feed grown and the manure-management systems change the scale of the impact and vary hugely from farm to farm but typical carbon emissions are comparable with those of dairy cattle.

Eggs labelled as 'organic' have lower emissions and soil and water impacts as the birds are fed an organically produced diet and the land used as outdoor pasture is rested for nine months between flocks, giving the vegetation and soil time to recover. Organic certification gives confidence that good animal welfare conditions are maintained because birds are free to roam outdoors and only given antibiotics when they are ill. There are harder challenges in manure management for flocks that spend time outdoors because rain can wash their phosphate-rich manure into water courses, triggering algal blooms.

Significant environmental impact is caused by egg waste. In 2018, in the UK alone 720 million good eggs were wasted,[102] in part because they are graded to be sold in single-size boxes and people discard them following the best-before dates on the box. To avoid this waste, buy mixed size eggs and try the simple water test to tell you which eggs are still perfectly good to eat: a healthy egg will sink to the bottom of a glass of water – on its side it is super fresh, on its end and it's less fresh but still fine to use – you may want to make sure it is cooked rather than using it raw. An egg that floats to the surface has to go out – floating indicates that gases from bad bacteria in the egg are building up and could make you sick.

Another tip: if you don't think you will get through your supply of eggs before they are past their best, they freeze well. Simply beat until just blended and put them in a freezer container – best to record the date and how many you have in there; they can be frozen for up to a year. When you want to use them, defrost in the refrigerator overnight and use the next day in dishes that are to be thoroughly cooked.

Can I eat any meat,
given climate concerns?

ISSUE

Most climate activists say we need to stop eating meat – around
14.5 per cent of our global greenhouse gas emissions comes
from livestock; roughly equivalent to global transport emissions.[103]
Is eating any type of meat OK?

SOLUTION The biggest problems with meat production are down to the effects of the scale and intensity of beef farming on some of the world's most sensitive and important ecosystems. Biodiverse rainforest and prairie are destroyed to provide grazing and to grow energy-rich soya-fed to cattle for faster weight gain to meet the demand for beef. The system is not only highly inefficient (the soya could feed people and would provide more direct energy and nutrition than being converted to meat), it causes greenhouse gas emissions (ruminants produce large quantities of methane), is pushing native species towards extinction and destroying the planet's ability to remove carbon from the atmosphere.

While red meat is a great source of protein, iron and vitamin B12, these nutrients can all be provided in well-managed vegetarian or vegan diets and too much meat can increase blood cholesterol and the risk of bowel cancer.[104, 105] Most of the western world needs to reduce meat consumption by around 50 per cent to meet Paris Agreement targets.[106]

Nevertheless, low-intensity meat production can be managed sustainably with lower emissions. Grazers are part of a healthy ecosystem, maintain open habitats for other wildlife, improving soil quality with their manure, using land unsuited to other purposes, and sustaining rural livelihoods and communities. It may not be necessary to stop eating meat but eat less and focus on where and it is produced. Buy from local suppliers who source meat from low-intensity, happy, healthy, grass-fed animals.

· ·

Good to know . . .

• Globally the total amount of food produced is enough for 5,940 kcal per person per day – roughly 2.5 times more than the average person needs to stay healthy. Of this, 1,740 kcal of human-grade food goes to feed animals, of which only 590 kcal is converted to meat and dairy and eaten (animals ingest an additional 3,810 kcal of grass and pasture).

• If the UK population adopted a vegetarian or vegan diet, the saving would equate to a 50% reduction in all UK car exhaust emissions.[107]

Does dairy have a big impact on the environment?

ISSUE

Milk and dairy products are produced from cattle, sheep, goats, camels and buffaloes, depending on where in the world those animals thrive. Does dairy farming cause the same problems of greenhouse gas emissions and deforestation as intensive meat production?

SOLUTION Again it's the type of dairy farming that largely dictates the scale of the problem. Large, intensively farmed herds with high stocking rates and extensive use of fertilizers and pesticides impact most. Like beef cattle, dairy cows emit methane and clearing land for grazing and feed can result in loss of sensitive habitats. However, dairy herds do not need to grow muscle fast and rely less heavily on soya-based feed, so dairy-associated emissions and deforestation are much lower, around a third of beef production.[108]

The impact of other milk-producing animals is lower than cows. Goats and camels are not full ruminants, and produce far less methane than cattle, although large-scale introduction of camels in Australia and goats in Europe could decimate ecosystems through overgrazing.

Plant-based milks produce just a third of the greenhouse gases and use a ninth of the land required to make a typical dairy milk. However, only soya has an equivalent nutritional value to animal milk. Soya-bean and coconut plantations replace biodiverse forest in the tropics, almond groves rely on copious irrigation and pesticides, which affect pollinators, and rice has the highest greenhouse gas emissions of any staple crop. Hazelnut and oat milks have probably the lowest impact if you want to be dairy-free.

Dairy, like beef, herds can form part of a sustainable biodiverse rich ecosystem if they are sustainably managed. Certification of organic milk by recognized bodies helps to identify the most environmentally sound products, typically from small producers using practices that allow herds to graze on grass and meadowland, that are fed only grass mixes, with their manure used positively to fertilize crops and produce energy.[109] It's a great example of a circular system (see page 106).

Dairy production is rising as diets become westernized – **6 billion** people worldwide consume dairy.

India is now the biggest producer of dairy (**20%** of global milk production) and the USA is the biggest producer of cow's milk (**12%** of global milk production).

In Europe, **84%** of dairy products are from farming systems that impact most on the environment; just **6%** come from ecologically valuable systems.[110]

Can I eat fish sustainably?

ISSUE

Fish seems like a healthy source of protein with a low carbon footprint but I also hear about overfishing, species decline and that factory ships are trawling out the seabed and creating a massive by-catch. Can I eat fish sustainably, ethically and without such waste?

SOLUTION Wild fish do have a low carbon fin-print, which suggests they could be a good source of essential protein without damaging the planet. Unfortunately, many of the world's fish stocks are fully exploited or overexploited and half of the global catch comes from industrial trawlers that engage in dubious ethical practices – trading at sea to avoid species quota restrictions, destroying the seabed and exploiting their onboard staff. The other, more sustainable, half of the global catch is from small-scale fishermen whose catch is providing essential nutrition to people in poorer countries of the world.

Farmed fish as an alternative has many of the environmental and ethical problems associated with farmed meat: the transfer of energy and nutrients from the feed they are given to the meat we eat is poor, overcrowding leads to sub-standard animal welfare resulting in high usage of antibiotics, and the concentration of organic-rich waste kills off local biodiversity – polluting entire ocean bays and destroying the seabed.

The ideal, if you want to eat fish, is to stick to sustainably caught wild fish – it can be a tricky balancing act. As with meat, consider fish as an occasional treat and try to find out if your fish is sustainably fished. The MSC (Marine Stewardship Council) is an international non-profit which certifies sustainable fisheries – each MSC-certified fishery has been independently assessed on its specific impacts to wild fish populations and the ecosystems they form part of. Where you see the blue fish logo you should be confident your fish has been sustainably sourced.

Choose pole-and-line caught fish, a method that will limit by-catches generally and of dolphin in particular, and avoid air-freighted fish for its carbon footprint. Fish that has been frozen and sent by ship is better than air-freighted if it has to come from the other side of the world. Buy from a fishmonger who can suggest less-common but tasty fish to replace the most commonly over-exploited fish species.

Should I buy local food?

I want to reduce my environmental impact when buying food. When I look at the origin of fruit and veg I realize some of it has travelled half way round the world, and I can't tell if it's come by air or ship. Is buying local produce always the best choice, and how local is local?

SOLUTION The transportation of food is usually only quite a small contributor to its overall carbon footprint, unless it is air-freighted. Buying local is therefore not always the best way to reduce its environmental impact. The majority of food-related greenhouse gas emissions reflect how the food was produced; for example, locally grown tomatoes in heated polytunnels in winter have much higher emissions than those shipped to your stores and markets from far sunnier countries where conditions are more suitable to growing and ripening the crop.[111]

With our global economy we have come to expect a wide range of fresh produce all year round. Overland transport is much more carbon efficient than air-freight so check the country of origin of fresh produce you want to buy and ask yourself if it would survive being transported by boat, train or road. Look closely at the origin of produce with a short shelf-life, such as grapes, pineapples, avocados and out-of-season strawberries or asparagus; they will likely have been flown thousands of miles – at great cost to the planet.

In season, locally grown produce is definitely the best option in terms of carbon footprint and there are knock-on sustainability and food-quality benefits in supporting farmers in your area too. Throughout the world, growing produce for the local community supports self-sufficiency, building in vital resilience to natural disasters and fluctuations in global fuel prices. Also, fresh produce that travels long distances quickly loses is flavour and nutritional value as it is picked before it is fully ripe, it requires packaging (frequently plastic) to extend its shelf-life and is often preserved by various means to survive the journey.

Whether you are at home or on holiday, there are local shops, markets and producers for your food needs. Hunt out these options, enjoy locally produced seasonal produce and a sense of community. And don't forget to see what you can grow yourself if you have some space (see page 56). All these options will have significant benefits for the environment as well as your personal health.

Which foods are responsible for the worst deforestation?

Our food choices have a huge impact on habitat loss – around 75 per cent of global deforestation is driven by food production.[112] Peatlands and rainforests are destroyed, driving species to extinction. I want to know which foods are responsible for the worst deforestation.

SOLUTION A handful of foods are the main drivers of deforestation globally. Forests are cleared for palm oil, coffee and cocoa plantations, while intensive farming practices mean trees are replaced by soya beans for fish and animal feed.

Palm oil is the world's most-traded vegetable oil. It's in everything from lipstick to biscuits, chocolate to laundry detergents. It is a very efficient crop, supplying 35 per cent of global vegetable oil needs on 10 per cent of the land other oil-producing plants require to produce that amount, and has the potential to be sustainably cropped.[113] The issue is, irresponsible production has devastated tropical peatlands and rainforests. There are websites and apps that identify palm oil products from sustainable sources, making it easy for you to favour retailers, manufacturers and brands that are committed to ensuring their products are responsibly produced.[114]

Our love of coffee and chocolate has driven the growth of these industries, too. The shade-loving coffee plant grows in rainforests, but by breeding sun-tolerant varieties, plantations now more than triple the crop, which has led to widespread deforestation in South America. Cocoa-bean production has caused similar deforestation in West Africa, Indonesia and the Amazon. Certification allows us to identify sustainably grown coffee and cocoa beans – look for Fairtrade, Organic and Rainforest Alliance marks on packaging and check your takeaway shop uses sustainably sourced coffee.

Weight for weight, soya beans provide more protein than meat. The deforestation problem arises from the vast quantities of soya used as animal feed: 80 per cent of the crop is fed to intensively farmed fish, cattle and chicken.[115] It's hugely inefficient: only a tenth of the weight is returned in the form of meat [116] and soya bean production is concentrated in some of the most biodiverse regions of China and South America.

Our food-buying habits empower us – you can simply avoid deforestation-driving products and remember that food produced sustainably works with nature and communities. Challenge shops and eateries to show that they use sustainable sources and check for certifications and badges to reward those that act responsibly.

How can I eat out and not impact the environment?

ISSUE

When we eat at home, we can buy locally from short supply chains and make sustainable choices. We can buy only what we need, use up or freeze any surplus and avoid products that drive deforestation. How can I enjoy eating out but ensure it is not having a huge impact?

SOLUTION Eating out is one of life's treats and should be savoured. Restaurants are becoming aware of their environmental impact and choosing well can lead to a guilt-free experience. There are things to look out for to ensure you choose environmentally friendly restaurants; if you make your favourite places aware of this they can help customers make sustainable choices by detailing on menus and marketing their supply chains, how they reduce food and non-food waste and how they manage their carbon footprint.

Food waste is a huge issue, so choose restaurants that put the right portion on your plate or offer self-service. If any food is left over, are there recyclable boxes so that you can take it home? Restaurants are often involved in schemes supporting food banks and homeless shelters. If not, ask them; it may encourage more thoughtful practice. Look at non-food waste too. Takeaways cannot eliminate packaging but is it minimal, non-plastic, fully recyclable or even returnable for them to reuse? Do they unthinkingly include napkins you probably won't use and sauces in single-use plastic sachets? All these add to the non-food waste burden of an outlet and are easily resolved.

Vegetarian and vegan options should look as good as meat and dairy ones. It may seem obvious but some restaurants still present one vegetarian or vegan dish as a token gesture and their appearance is distinctly second best. The range of plant-based meals available when eating out is expanding and increasingly more inviting, which encourages more sustainable diets. Use apps and websites to help you find sustainable eating options, post reviews and talk about your own discoveries on social media to grow the network of environmentally friendly places to eat.

In the UK, hospitality and food service annually produces **2.87 million tonnes** of food and non-food waste. Of the **1 million tonnes** that is food, **75%** could have been eaten.

45% of food waste is from food preparation and **21%** from spoilage.

1.87 million tonnes is packaging and non-food waste, **56%** of which could be recycled.[117]

It has been estimated that the annual supply of single-use sauce sachets, laid out flat, would cover the globe.[118]

How can I reduce plastics in my food shop?

Plastics are derived from fossil fuels, require more fossil fuels to manufacture, and release pollutants into the atmosphere. They last for centuries; only 9 per cent is recycled and some 8 million tonnes end up in our waterways annually. How can we cut this pollution?

SOLUTION The best option is to shop in plastic-free and refill shops or markets where you can reuse containers. But even if your only option is the local supermarket there are some tricks to cut down on your plastic load.

Take your own reusable bags or boxes to transport your shopping home and take your own reusable bags for produce, use paper bags or think about whether you need a bag at all – bananas, pineapples, squash and onions are ready wrapped by nature – do they need to be bagged? Avoid fresh produce that comes bulk-buy in bags or wrapped in plastic – this way you not only save on plastic but probably only buy what you need. If you forget your bags, ask for a cardboard box rather than buying a 'bag-for-life'.[119]

More supermarkets are opening bulk food sections where you can bring your own containers, so make sure you use them when they are available. Meat, fish and cheese counters at most supermarkets will let you use your own containers; do ask if it's not advertised.

Pick bottled goods in glass rather than plastic and make sure you reuse and recycle. Drinks cartons are a better choice than plastic bottles (see page 26) – check that you can recycle them in your area.

Grow your own – many herbs that are expensive and come in plastic pots or bags can be grown from seed on your own windowsill for a fraction of the cost and lots of satisfaction (see page 56). And, sometimes, if there is no alternative, be prepared to skip a product that just has too much plastic – make sure you tell someone at customer services or online – they may change their ways to keep your custom.

All these are small steps in the right direction. Be kind to yourself; while we want to make progress, it can be hard to achieve a fully plastic-free shop if your only choice is a supermarket. Try to make sure that any plastic you end up with is recycled – either in your kerb-side recycling, specialist recycling collections or back at the supermarket.

We see food waste and food poverty: what's the answer?

ISSUE

The statistics are alarming: 821 million people worldwide go hungry, while 1.9 million are obese.[120] In the EU 55 million people can't afford a decent meal each day while 20 per cent of food produced is wasted. Our food system is failing us and the planet. What can I do to help?

SOLUTION   Food poverty is caused by a number of issues that are often exacerbated by crises such as pest outbreaks, adverse weather conditions and wars and political instability in some regions and, increasingly, natural disasters driven by climate change. Globally, the growing numbers of displaced people will continue to drive appalling bad food poverty statistics. However, even in some of the world's richest countries, short-term crises in finances or personal circumstances, often hidden behind closed doors as well as on our streets, are resulting in food poverty: people are hungry and malnourished. In the UK for example, in 2016, the largest food bank trust provided 1.2 million packages of emergency food supplies. Our actions locally can help reduce food waste and support a fairer distribution of food.

We have personal responsibilities, which are covered elsewhere in this chapter, to show how we can ensure we waste as little food as possible at home and that our supply chains are equally well managed and sustainable. It's important to get the issue in the public eye: we can lobby our governments and local councils for better policies, plans and strategies to tackle the root causes of food poverty and find solutions to help families out of trouble. We can also lend our support to charities that are on the front line of supporting people who are struggling to feed themselves and their families. We and the food outlets we use can support by donating food, money and by volunteering our time at food banks, homeless shelters and breakfast clubs. Apps that link charities with retailers and restaurants make it even easier.

Finally, on a one-to-one basis, we can also be aware of vulnerable individuals such as young families, the elderly, or house-bound neighbours who need a hand with food shopping and cooking to ensure they do not become malnourished.

How can I reduce food waste down the food supply chain?

ISSUE

Tackling food waste is the third most effective solution to tackling climate change[121] and waste is apparent all the way down the supply chain, from field to fork. Around two-thirds of food production is lost in harvesting and storage; what can I do to minimize this waste?

SOLUTION As consumers the best way that we can reduce food waste in the food supply chain is to change our buying patterns. Much of the waste is driven by supermarket buying practices. Supermarkets themselves waste very little food; after all, food is their profit. Once produce reaches the supermarkets only around 0.5 per cent of it is wasted. However, supermarkets can drive huge amounts of waste on farms. Faced with inflexible contracts from supermarkets, farmers overproduce to allow for losses from increasingly unpredictable weather or pests to ensure they are able to fulfil their orders, and the excess is often wasted. Similarly, the demand for uniform shapes, sizes and cosmetic standards means many crops are rejected and wasted before reaching the shelves.

There is wastage at every stage of the complex supply chain that food undergoes: harvesting—storage—transportation—refrigeration—transportation—storage—processing—storage—transportation.

The best way that we can reduce waste in the supply chain is to eliminate as many of these links as possible, reducing the length of the chain. If we can get the food from field to consumer in as few stages as possible, we cut out waste at every stage. This means that your best bet may not be visiting the supermarket, which may mean accepting less choice, and instead opting for locally and seasonally produced fruit and veg. Local farm shops, market stalls and box schemes usually have shorter supply chains and often do their best to eliminate wastage. Find out from your farm supplier what they do with food they can't sell. They should give it to their staff, use it in staff canteens, give it away to food banks or feed it to livestock.

Being well informed about supply chains is one of the most reliable ways we can reduce our impact on the planet. It may take some time, commitment and energy but once you find suppliers you trust, you will be rewarded with healthier, more nutritious food that is not costing the earth.

How can I reduce the impact of buying things online?

The carbon cost of delivering online purchases is huge. Packages may be shipped long distance to reach the distribution point, while the 'last mile', from depot to doorstep, can account for up to 50 per cent of the carbon and air-polluting emissions of the product's entire journey.[122]

SOLUTION Before you decide never to shop online again remember that the carbon cost of delivery may still be less than you making a solo car journey to get one or two items. Logistics companies are devising innovative ways to reduce the impact of the 'last mile', including delivery hubs in garages and local shops, using existing services such as refuse collection vans or delivery safe boxes; even access to your car boot.[123] However, until these options are rolled out more widely there are still ways you can cut the carbon cost of that 'last mile'.

Choose standard delivery whenever possible. Last-minute shopping and rushed orders for next-day delivery require logistics companies to lay on a special service to get your product to you, rather than having time to consolidate their journeys and combine deliveries more efficiently in the way that supermarkets organize delivery of groceries. Don't miss deliveries. Taking products away and having to redeliver them doubles the already large carbon footprint of the 'last mile'. If you know you will be out, arrange a safe place where your parcel can be left, either on your property or with a neighbour.

Buy carefully so you don't have to return products: this is really bad for the environment. Returning an item involves a repeat logistics chain back to the retailer and, worse, the challenges of processing that return and getting it in condition for resale can often be so great for the retailer that a lot of returned items (particularly clothes) just end up in landfill.

Rather than randomly buying throughout the month, try to set aside one day when you do all your online shopping and get everything in one go, preferably from one online retailer. This saves on delivery journeys and also reduces packaging.

And finally – buy less stuff. Before you hit that button, ask yourself: *Do I need it and do I really like it, will I use it and appreciate it?* – or will it be rejoining the logistics delivery chain when you sell it on or even going to landfill when you change your mind in a month or so?

Conclusion: what we have discovered

The topics in this book demonstrate the scale and complexity of the task we face to live an eco-friendly lifestyle. The answers to our daily questions are not always clear cut. What at first seems like a solution to solve a problem becomes more nuanced and opaque as we dig deeper. Some solutions have knock-on effects and create further problems. The global nature of problems can also seem overwhelming but there are small, daily actions we can all take that have a big impact on the planet. This book is a starting point for ideas that will help you take some control and make a contribution to solving the challenges of climate change and loss of biodiversity. Fresh analysis and changing human needs regularly reveal shifting priorities and developing technology provides options for better solutions. Keep reading and listening and talking to stay informed. Search out the best information.

What comes next? World leaders are finally, properly, waking up to the environmental challenges. At their annual gathering in Davos 2020, politicians and business leaders acknowledged that the top five risks to the future global economy are all climate- and biodiversity-related; this is a huge step in moving towards finding global solutions. Let us hope our leaders speak as one when they gather at UN assemblies and conventions to renew the agreements on achieving global net zero emissions and saving the planet's biodiversity.

Whatever happens through political resolution, there is cause for optimism and we all can play a role. We are small, but we are many and we share one home. Given the space, nature has an amazing capacity for regeneration, and we need to work together to make that happen.

It is important to be alert to the most reliable information as we change our habits and behaviour to help us solve our present-day problems. In a changing world our automatic responses may no longer be appropriate or productive. The best way to combat this is to stay informed, through the most reliable sources. In unsettled, often anxiety-inducing, times, we need to be discerning and keep probing, maintain an open mind and a flexible outlook to respond with actions appropriate to the developing situation. Continue to question and challenge leaders, service providers and the media; demand the truth. If you believe you have found the right answer to a particular question, then tell others about it. Be a strategic consumer and keep spreading the word; pass on your thinking to others so that we can all make informed choices. We can't always know the best answer but don't be paralysed by imperfect science and be prepared to keep reconsidering. If all else fails stick to these simple rules:

• Use less and enjoy it more
• Find out about supply chains and support ones that are environmentally aware
• Use the option with the smallest carbon footprint
• Choose the option that results in the least waste, and make choices that support your local community and enable the natural world to be more resilient.

Finally, keep going. Our anxiety stems from what neuroscientists term learned helplessness. It develops when we desperately want to see change but find it hard to have an impact; we see our actions as ineffective, or that change is neither quick nor visible enough. If we allow ourselves to get demoralized and develop a sense of helplessness, anxiety takes over and we cease to act. None of us has the solution to our ailing world, but together we can make big changes occur: we are all part of the whole. So live in the now and celebrate the local; while having an eye to the future and considering the global.

References

Indoors

1. www.betterhealth.vic.gov.au/health/conditionsandtreatments/antibacterial-cleaning-products

2. noharm-uscanada.org/issues/us-canada/cleaners-and-disinfectants

3. www.coolaustralia.org/wp-content/uploads/2012/12/Water-fact-sheet.pdf

4. www.sciencedaily.com/releases/2016/03/160304092230.htm

5. ec.europa.eu/environment/chemicals/endocrine/definitions/affect_en.htm

6. www.bpf.co.uk/packaging/why-do-we-need-plastic-packaging.aspx

7. www.independent.co.uk/life-style/plastic-bad-environment-why-ocean-pollution-how-much-single-use-facts-recycling-a8309311.html

8. BP (2018) statistical review of world energy 2018 www.bp.com/content/dam/bp/business-sites/en/global/corporate/pdfs/energy-economics/statistical-review/bp-stats-review-2018-full-report.pdf

9. Berners-Lee, M. (2019) *There is No Planet B*, page 60.

10. Captain calculator calories burned during sleep calculator – sleep calories captaincalculator.com/health/calorie/calories-burned-sleeping-calculator/

11. Berners-Lee, M. (2019) *There is No Planet B*, page 60.

12. theconversation.com/sustainable-shopping-how-to-stay-green-when-buying-white-goods-89454

13. www.smartenergygb.org/en/resources/press-centre/press-releases-folder/usage-tracker-may-2019

14. www.woodmac.com/news/editorial/global-smart-meter-total-h1-2019/

15. smartenergygb.org

16. www.energuide.be/en/questions-answers/do-i-emit-co2-when-i-surf-the-internet/69/

17. www.capgemini.com/gb-en/wp-content/uploads/sites/3/2017/07/crs_stories_to_be_proud_of_the_merlin_data_centre_final_19-10-15.pdf; www.nature.com/articles/d41586-018-06610-y;orleansmarketing.com/35-technology-facts-stats/

18. Dela Cruz M. (2014) 'Can ornamental potted plants remove volatile organic compounds from indoor air? — a review.' *Environ. Sci Pollut Res.* doi: 10.1007/s11356-014-3240-x

19. Caulfield T. (2015) 'The Pseudoscience of Beauty Products', *The Atlantic*, 5 May 5 2015.

20. www.bbc.co.uk/news/science-environment-41570540 1

21. www.ellenmacarthurfoundation.org/publications/a-new-textiles-economy-redesigning-fashions-future

22. www.vox.com/the-goods/2018/9/19/17800654/clothes-plastic-pollution-polyester-washing-machine

23. ecocult.com/greenwashing-alert-that-natural-fabric-made-from-plants-might-be-toxic/

24. finisterre.com/blogs/fabric-of-finisterre/organic-cotton

25. www.soilassociation.org/organic-living/fashion-textiles/organic-cotton/

26. goodonyou.eco/most-sustainable-fabrics/

27. www.ethicalconsumer.org/fashion-clothing

28. www.worldwildlife.org/industries/cotton

29. www.theatlantic.com/science/archive/2019/02/deepest-ocean-trenches-animals-eat-plastic/583657/

30. Hartline N. et al. (2016) 'Microfiber masses recovered from conventional machine washing of new or aged garments', *Environ. Sci. Technol.* 50:21,11532–8. doi/abs/10.1021/acs.est.6b03045?journalCode=esthag#

Outdoors

31. www.iucn.org/resources/issues-briefs/peatlands-and-climate-change

32. IUCN Resolution on Global Peatlands portals. iucn.org/library/sites/library/files/resrecfiles/WCC_2016_RES_043_EN.pdf

33. Sanchez-Bayo F. and Wyckhuys K. (2019) 'Worldwide decline of the entomofauna: A review of its drivers'. *Biol. Cons.*, 232: 8–27. doi.org/10.1016/j.biocon.2019.01.020

34. www.birdlife.org/worldwide/news/new-study-conservation-action-has-reduced-bird-extinction-rates-40

35. Thompson, K. and Head, S. 'Gardens as a resource for wildlife' http://www.wlgf.org/The%20garden%20Resource.pdf

36. time.com/5539942/green-space-health-wellness/

37. www.nhs.uk/conditions/stress-anxiety-depression/improve-mental-wellbeing/

38. RHS (2019) Gardening without harmful invasive plants. http://www.nonnativespecies.org/index.cfm?pageid=303

39. USA - www.nwf.org/Educational-Resources/Wildlife-Guide/Threats-to-Wildlife/Invasive-Species

40. www.ipcc.ch

41. Bastin, J.-F. (2019) 'The global tree restoration potential', *Science* 365: 448:76–9 doi: 10.1126/science.aax0848

42. www.woodlandtrust.org.uk/plant-trees/advice/grow-from-seed/

43. www.gardeningknowhow.com/composting/basics/making-compost-indoors.htm

44. www.organicauthority.com/live-grow/best-herbs-to-grow-for-windowsill-herb-garden

45. www.gardeners.com/how-to/grow-microgreens/7987.html

46. http://www.eatingwell.com/article/290729/how-to-grow-fruits-vegetables-from-food-scraps/

47. time.com/5539942/green-space-health-wellness/

48. lnt.org/why/7-principles/

49. www.rspca.org.uk/adviceandwelfare/litter

50. www.abc.net.au/news/2014-09-15/plastic-pollution-choking-australian-waters-study/5744398

51. litteritcostsyou.org/9-interesting-facts-and-statistics-about-littering/

52. doodycalls.com/blog/epa-says-dog-poop-is-an-environmental-hazard-on-par-with-pesticides/

53. www.countryfile.com/wildlife/mammals/neosporosis-the-hidden-danger-that-dogs-pose-to-cattle/

54. pfa.org.uk

55. www.envirotech-online.com/news/environmental-laboratory/7/breaking-news/how-long-do-biodegradable-bags-really-last/49246

56. www.outsideonline.com/2292736/its-time-talk-about-dog-poop#close

Transport

57. BBC 'Climate change: should you fly drive or take the train?' www.bbc.co.uk/news/science-environment-49349566 57

58. www.c2es.org/content/reducing-your-transportation-footprint/

59. www.gov.uk/government/publications/greenhouse-gas-reporting-conversion-factors-2019

60. theicct.org/sites/default/files/publications/ICCT_CO2-commercl-aviation-2018_20190918.pdf

61. www.goldstandard.org

62. unfccc.int/climate-action/climate-neutral-now

63. www.euronews.com/2019/08/17/exporting-contamination-who-pays-the-environmental-cost-of-electric-car-production

64. www.mining.com/tesla-warns-upcoming-battery-minerals-shortage/

65. www.medicalnewstoday.com/articles/324483.php

66. Royal College of Physicians (2016) tinyurl.com/pollutiondiesel / PHE (2014) tinyurl.com/deathsdiesel

67. Berners-Lee, M. (2019), page 106.

68. www.tmr.qld.gov.au/Travel-and-transport/Cycling/Benefits.aspx

69. www.who.int/dietphysicalactivity/factsheet_adults/en/

70. www.fueleconomy.gov/feg/maintain.jsp

71. www.ornl.gov/news/sensible-driving-saves-more-gas-drivers-think

72. https://www.ornl.gov/publication/fuel-economy-and-emissions-effects-low-tire-pressure-open-windows-roof-top-and-hitch. Study results are based on testing with a small sedan, a standard size SUV, a single roof-top cargo box (20" h x 40" w x 50" l), and a single rear-mount cargo tray. Cargo boxes with other dimensions or shapes may have a different effect on fuel economy.

On Holiday

73. www.europarl.europa.eu/RegData/etudes/STUD/2018/629184/IPOL_STU(2018)629184_EN.pdf

74. http://haereticus-lab.org/most-sunscreens-can-harm-coral-reefs-what-should-travelers-do/

75. http://haereticus-lab.org/protect-land-sea-certification-3/#link-target1/

76. www.e-unwto.org/doi/pdf/10.18111/9789284421152

77. www.airbnb.co.uk/press/news/new-study-reveals-a-greener-way-to-travel-airbnb-community-shows-environmental-benefits-of-home-sharing

At Work

78. www.conserve-energy-future.com/real-figures-paper-usage-uk.php

79. www.statista.com/topics/1701/paper-industry/

80. www.statista.com/statistics/1033512/recycling-rate-of-paper-and-cardboard-packaging-waste-in-the-eu-by-country/

81. Suez (2019) Paper and cardboard recycling and recovery factsheet www.suez.com.au/en-au/sustainability-tips/recycling-fact-sheets

82. Cambridge Architectural Research (Nov. 2012) 'How much energy could be saved by making small changes to everyday household behaviours?' assets.publishing.service.gov.uk/government/uploads/system/uploads/attachment_data/file/128720/6923-how-much-energy-could-be-saved-by-making-small-cha.pdf

83. news.cornell.edu/stories/2004/10/warm-offices-linked-fewer-typing-errors-higher-productivity

84. www.pwc.co.uk/who-we-are/our-purpose/low-carbon-circular-business/waste/give-me-tap.html

85. Kenefick R. and Sawka M. (2007) 'Hydration at the work site', *J. Am. Coll. Nutrition* (26: 5, 597S–603S) www.researchgate.net/profile/Robert_Kenefick/publication/5924534_Hydration_at_the_Work_Site/links/54ad606a0cf2213c5fe3d3cb/Hydration-at-the-Work-Site.pdf

86. Tine, M. T. and Butler, A. G. (2012) Acute aerobic exercise impacts selective attention: an exceptional boost in lower-income children, *Edu Psych.*, 32:7, 821–34 doi: 10.1080/01443410.2012.723612

87. www.businessleader.co.uk/how-long-is-the-daily-work-commute-for-the-average-londoner/44325/

88. www.cnbc.com/2018/02/22/study-states-with-the-longest-and-shortest-commutes.html

89. www.abs.gov.au/ausstats/abs@.nsf/Lookup/by%20Subject/2071.0.55.001~2016~Main%20Features~Feature%20Article:%20Journey%20to%20Work%20in%20Australia~40

90. David Attenborough www.ourplanet.com/en/video/a-reason-for-hope/, wwf.panda.org/our_work/projects/our_planet_netflix_wwf_nature_documentary/ and www.youtube.com/watch?v=YAkIsbrRKWU

91. Greta Thunberg (Davos) www.youtube.com/watch?v=KAJsdgTPJpU and www.youtube.com/watch?v=11FCyUB81rI

92. www.pwc.co.uk/who-we-are/our-purpose/low-carbon-circular-business/waste/give-me-tap.html

93. friendsoftheearth.uk/climate-change/over-twothirds-young-people-experience-ecoanxiety-friends-earth-launch-campaign-turn

94. yougov.co.uk/topics/politics/explore/issue/Climate_change

95. www.inaturalist.org/

96. ebird.org/home

97. www.zooniverse.org

98. www.circularity-gap.world

99. www.bitc.org.uk/wp-content/uploads/2019/10/bitc-globalgoals-toolkit-owntheconversationkit-sept2019.pdf

100. www.ellenmacarthurfoundation.org/circular-economy/concept

Food & Shopping

101. Recipes for leftovers. www.lovefoodhatewaste.com

102. www.wrap.org.uk/food-drink WRAP is the UK government advisers on food waste

103. www.cbc.ca/news/technology/livestock-ghg-emissions-science-1.4753165

104. www.nhs.uk/live-well/eat-well/meat-nutrition/

105. www.health.harvard.edu/staying-healthy/whats-the-beef-with-red-meat

106. Berners-Lee, M., Hoolohan C., Cammack C. and Hewitt C.N. (2011) 'The relative greenhouse gas impacts of realistic dietary choices'. doi.org/10.1016/j.enpol.2011.12.054

107. ibid.

108. Berners-Lee, M. (2019) *There is No Planet B*, page 24.

109. www.ecoandbeyond.co/articles/sustainable-dairy-farms/

110. ec.europa.eu/environment/agriculture/pdf/dairy_xs.pdf

111. Watkiss P. et al. (2005) 'The validity of food miles as an indicator of sustainable development'. AEA technology environment study for Defra

112. www.onegreenplanet.org/environment/foods-that-are-eating-the-worlds-forests-and-how-to-choose-better/

113. www.wwf.org.uk/updates/8-things-know-about-palm-oil

114. gikibadges.com and WWF scorecard palmoilscorecard.panda.org

115. wwf.panda.org/our_work/food/sustainable_production/soy.cfm

116. Berners-Lee, M. (2019) *There is No Planet B*, page 17.

117. www.vice.com/en_uk/article/8xyvpb/all-the-ways-restaurants-ruin-the-environment

118. inews.co.uk/news/environment/plastic-pollution-sauce-sachets-ketchup-mayonnaise-landfill-recycling-1891543

119. www.self.com/story/how-to-cut-down-on-single-use-plastic-grocery-shopping

120. wwf.panda.org/our_work/food/food_loss_and_waste/

121. Riverford - choose.riverford.co.uk/say-no-to-food-waste/?utm_campaign=Customer+Active+Choose+Food+Waste+13+09+2019&utm_medium=email&utm_source=Master+Riverford+Customers

122. www.greenlogistics.org/SiteResources/dcc2b4e4-52ef-4fa1-8ad9-f96a7f43f976_JE-AMcK-LRN%20-%20Last%20Mile%20-%20Presentation.pdf

123. Buldeo Rai, H. et al. (2019), 'The "next day, free delivery" myth unravelled: Possibilities for sustainable last mile transport in an omnichannel environment', *Int. J. Retail & Distribution Management*, 47 (1): 39–54. doi.org/10.1108/IJRDM-06-2018-0104

Further Reading & Resources

Books & Useful Websites

Berners-Lee, M. (2011) *How Bad are Bananas? The carbon footprint of everything* (Green Profile)

Berners Lee, M. (2019) *There is No Planet B: A handbook for the make or break years* (Cambridge)

Bridle, J. (2019) *New Dark Age – Technology and the end of the future* (Verso)

Browne, M. et al. (2019) *Urban Logistics: Management, policy and innovation in a rapidly changing environment* (Kogan Page)

Carson, Rachel (1962) *Silent Spring* (reissued 2000 Penguin)

Gore, Al (2006) *An Inconvenient Truth* (Rodale)

Hawken, Paul (2018) *Drawdown: The most comprehensive plan ever proposed to reverse global warming* (Penguin)

Thunberg, Greta (2019) *No One is Too Small to Make a Difference* (Penguin)

Tree, I. (2019) *Wilding: The Return of Nature to a British Farm* (Picador)

At Home & Gardening

www.bettercotton.org
www.ecover.com
www.energysavingtrust.org.uk
www.koh.com
www.methodhome.com
www.rspb.org
www.ethicalconsumer.org
www.buglife.org
www.wildaboutgardens.org

Food & Shopping

www.eatlowcarbon.org
www.fairtrade.org.uk
www.food.cloud
www.goodfish.org.au

Marine Council Stewardship see in particular www.mcsuk.org/goodfishguide/search

www.olioex.com – food sharing

www.rainforest-alliance.org

www.soilassociation.org

www.lovefoodhatewaste.org

Recycling & the Circular Economy

www.ellenmacarthurfoundation.org
www.terracycle.com
www.wrap.org.uk

Take Action

www.foe.org – friends of the earth
www.greenpeace.org.uk
www.carbontrust.com

Travel

www.responsibletravel.com
www.liftshare.org
www.energysavingtrust.org.uk/transport-travel

Wildlife & Conservation Organizations

www.arborday.org
www.natureaustralia.org.au
www.nature.org/en-us/
https://onetreeplanted.org
www.theriverstrust.org
www.teamtrees.org
www.wildlifetrusts.org
www.woodlandtrust.org.uk
www.worldwildlife.org

Index

Acknowledgements

This book has grown from the encouragement and enthusiasm from people close to me, people I work with and people I meet in passing. There is such concern about our planet that I have felt carried along by people's need for this book.

I would like to thank those closest to the project who really made it happen. To all the dedicated team at Leaping Hare Press – in particular Monica Perdoni who made the links and Niamh Jones, Stephanie Evans and Tom Kitch who brought it all together.

I have been inspired by so many but particularly the wonderful people I have come across in the Year of Green Action and the Council for Sustainable Business who lift my spirits with their commitment to green action and environmental sustainability and help me believe we really can make a difference – a particular shout out to the small but perfectly formed core Year of Green Action team for an amazing year: Helen, Tracey and Paul.

Life would not be the same without my friends who bring fun, laughter and light and most of all my family; the Turners – Welsh and Dorset – Foulshams and Wardleys who demonstrate the best of humanity and give me all the love and support I could ever wish for.

And even more than everyone else... the people who put up with me day to day – my gorgeous Dunc, wonderful Anousha, Poppy, Thea and Lottie and the love of all our lives Alfie.